John Morse Ordway

Analytical Guide to the Natural Orders of the Vegetable Kingdom

Printed for the use of students in the Massachusetts Institute of

Technology

John Morse Ordway

Analytical Guide to the Natural Orders of the Vegetable Kingdom
Printed for the use of students in the Massachusetts Institute of Technology

ISBN/EAN: 9783337165000

Printed in Europe, USA, Canada, Australia, Japan

Cover: Foto ©berggeist007 / pixelio.de

More available books at **www.hansebooks.com**

ANALYTICAL GUIDE

TO THE

NATURAL ORDERS

OF THE

VEGETABLE KINGDOM.

PRINTED FOR THE USE OF STUDENTS

IN THE

MASSACHUSETTS INSTITUTE OF TECHNOLOGY.

BOSTON:
W. J. SCHOFIELD, PRINTER, 105 SUMMER STREET.
1881.

NOTE.

The analytical tables usually found in descriptive botanical works are partial, and contain only such natural orders as are represented in one country, or even in a limited portion of a very large country like our own. But in these latter days of good traveling facilities few persons need confine their attention to a narrow district. Moreover, the frequent introduction of new plants for cultivation is extending the range of species that those who stay at home may desire to study. In teaching botany to those who are going to various parts of the world, the writer has felt the need of comprehensive tables which might serve as an index to the whole vegetable kingdom, — to the flowerless as well as the flowering plants. The want was, in a measure, supplied by printing in 1874, for his own classes, a table based on those contained in Balfour's Class Book of Botany. The form, however, was changed, since for the student who, with a plant in hand, seeks to determine its natural order the older form of conspectus is far less convenient than the plan of alternative distinctions arranged in pairs with onward references. Reference figures are simpler and less confusing than asterisks, daggers, and letters, and the method here introduced avoids the needless multiplication of numbers. But as Balfour's tables have, in use, proved too concise and not altogether exact, the whole matter has, with much labor, been revised, enlarged, and recast. A multitude of botanical terms which remain familiar to the specialist cannot be always borne in mind by the ordinary student, and, therefore, technical language has been dispensed with as far as possible, even somewhat at the expense of brevity and precision. Of course, one whose life work has not been in the special line of botany can expect but partial success in going over so extensive a field, and there must needs still remain many defects and perhaps errors. Endlicher's Genera Plantarum has proved of great service in the consideration of the phanerogamous plants. The table of cryptogamous plants has been made out by taking from various sources what seemed to be the best classifications given by special writers, there being no recent, single work which gives a full account of the orders in all departments.

The Filices, or Ferns, are arranged to suit Hooker and Baker's "Synopsis Filicum, 2d ed., London, 1874." The Musci are made to conform to Schimper's "Synopsis Muscorum Europæorum, 2d ed., Stuttgartiæ, 1876." But the tables of the Musci and the Hepaticæ are taken, in great part, from Debat's "Flore des Muscinées, Lyon, 1874." That of the Musci, however, has been revised, condensed, and modified to make it accord with the new edition of Schimper's great work, Debat having used the first edition. The orders Drepanophyllaceæ, Phyllogoniaceæ, Hypopterygiaceæ, including mosses found only between the tropics or in southern latitudes, are inserted from Müller's " Synopsis Muscorum, frondosorum omnium hucusque cognitorum, Berlin, 1849."

The fresh water Algæ are classified according to Rabenhorst's "Flora Algarum Aquæ Dulcis et Submarinæ, Lipsiæ, 1864–8." The marine Algæ are taken from Agardh's "Species et Ordines Algarum, Lundæ, 1848 — 76."

Leading authors differ much in the systematic arrangement of the Lichens, and a table cannot be made to suit more than one of the standard works. It was thought best to follow the latest, Tuckerman's "Genera Lichenum, Amherst, 1872." The Fungi are given partly in accordance with Cooke's "Handbook of British Fungi, London, 1871," as well as his "Fungi; their Nature and Uses, New York, 1875," though neither of these works is in full accordance with the present development of mycology.

The study of the genera and species of the cryptogamous plants is difficult, because some of the parts on which the distinctions are based are very minute, and require the use of the microscope; because the range of variation in the same species is often large; and because in many of them the organs of reproduction are rarely met with. The natural difficulties are still farther aggravated by the unnecessary multiplication of terms used by different specialists to designate similar organs. Indeed, the almost limitless resources of the Greek language have been severely taxed to supply words for successive authors, and there is, as yet, no single glossary to be found which gives a full explanation of all the necessary and the superfluous names. Another trouble yet is the want of books describing, in one work, the whole cryptogamic flora of one country.

For the student in the northern United States, Gray's Manual and Wood's Class Book give most of the Filices, Equisetaceæ, Lycopodiaceæ, Isoetaceæ, Marsiliaceæ, and Salviniaceæ. The Musci and Hepaticæ are described by Sullivant, whose work has been incorporated with Gray's Manual. Harvey's Nereis Boreali-Americana, and Wood's "Contribution to the History of the Fresh Water Algæ of North America, Washington, 1872," must be referred to for our Algæ; and for the Diatomaceæ we must go to the work of Rabenhorst, which is written in Latin and does not apply specially to our country. Tuckerman gives only generic descriptions of the Lichens. Cook's Handbook of British Fungi includes a large number of American species.

Of late years a multitude of zealous workers have cleared up much of the mystery which enshrouded the flowerless plants in the time of Linnæus, and the time is not far distant when it will be no longer correct to speak of the phanerogamic and cryptogamic, but the distinction will be between seed plants and spore plants.

Before proceeding with the artificial, analytical key we may give a concise, general view of the vegetable kingdom thus: —

I.—FLOWERING PLANTS.

A.— Exogens.

  A.— Perianth of two whorls or more.

    1.— Corollifloral.

      a.— Stamens perigynous.

      b.— Stamens hypogynous.

    2.— Calycifloral.   3.— Thalamifloral.

      a.— Polypetalous.   a.— Stamens few.

      b.— Monopetalous.   b.— Stamens many.

  B.— Perianth of one whorl or none.

    1.— Angiospermous, as Salicaceæ.

    2.— Gymnospermous, as Coniferæ, Cycadaceæ.

B.— Endogens.

    a.— Petaloideous, as Liliaceæ.

    b.— Spadiceous, as Palmæ, Araceæ.

    c.— Glumaceous; Cyperaceæ, Gramineæ.

II.—FLOWERLESS PLANTS.

A.—Vascular.

  a.—Spores of two kinds, or *heterosporous.*
    1.—Fruit on special stems; Salviniaceæ, Marsiliaccæ.
    2.—Fruit on the leaves; Isoetaceæ, Sclaginellaceæ.
  b.—Spores of one kind, or *isosporous.*
    1.—Fruit on special scales; Equisetaceæ, Lycopodiaccæ.
    2.—Fruit on the leaves; Filices.

B.—Cellular.

  a.—Acrogens.
    1.—Muscineæ; Musci, Hepaticæ.
  b.—Thallogens.
    1.—Carposporous.
      a.—Algous, or with chlorophyll; Rhodophyceæ.
      b.—Mixed; Lichens.
      c.—Fungous, or without chlorophyll.
        a.—Hymenomycetes, as Agaricini, Polyporci.
        b.—Gasteromycetes, as Lycoperdacei.
        c.—Tremellini.
        d.—Æcidiomycetes.
        e.—Ascomycetes.
    2.—Oosporous.
      a.—Algous, as Fucaceæ.
      b.—Fungous, as Peronosporeæ.
    3.—Zygosporous.
      a.—Algous.
        a.—With zoospores, as Ulotrichaceæ.
        b.—Conjugates, as Zygnemaceæ.
      b.—Fungous.
        a.—Myxomycetes.
        b.—Zygomycetes.
    4.—Protophytes.
      a.—Algous.
        a.—Chlorophyllophyceæ.
        b.—Cyanophyceæ.
      b.—Fungous.
        a.—Saccharomycetes.
        b.—Schizomycetes.

Plants are distinguished from animals by the absence of a stomach, a permanent or extemporized internal cavity for the reception of liquid or solid food. Though a very few plants have hollow leaves or utricles, or irritable leaf hairs which close around an insect lighting on them, such plants still depend, for their nourishment, mainly on inhaled carbonic acid and liquids absorbed by external roots. Plants are also chemically characterized by having their cell walls constituted of cellulose, a substance containing no nitrogen.

The following table is made up of alternative distinctions successively pointing out to the student which of two or more paths he must follow to determine the natural order of the specimen in hand.

If no name or number follows the right description, he should go directly on to the next set; if a number follows the correct description, he should turn to the reference given; and the comparisons should thus be continued till the name of the order is found.

The small figures after the name of a natural order indicate the page of Gray's "Field, Forest, and Garden Botany" on which that order is described. But sometimes Gray recognizes as sub-orders what others reckon as orders, and in such cases his description accompanies a different name from the one here used.

When no figures follow the name of a phanerogamous order, no species of the order is known to grow wild in the northern United States.

ANALYSIS.

1. — Flowerless plants, or those reproducing by mere cell division or by means of spores which usually consist of a single cell. **125**
Flowering plants, or those reproducing by seeds which are made up of numerous cells not all alike.

Parts of the flower generally in fives or twos, rarely in threes. Leaves net-veined. Fibrovascular bundles of the stem systematically arranged around a central pith. Stem with a distinct bark. Embryo or germ of the seed with a pair or whorl of cotyledons, or rarely with none. *Exogens.* **2**
Parts of the flower generally in threes, but sometimes in twos or fours. Leaves mostly, but not in all cases, parallel-veined. Fibrovascular bundles of the stem arranged around the outside, and scattered among the cellular tissue. With no true bark. Embryo with a single cotyledon. *Endogens.*

Perianth consisting of *glumes*, that is, of imbricated or overlapping chaff-like bracts, which are usually in pairs. Leaves always parallel-veined, long and narrow, often sheathing at base. **122**
Perianth not glumaceous, or altogether lacking. **109** ·

2. — Carpels entirely open, neither infolded nor mutually cohering by their edges. Seeds not contained in a closed pericarp, but either attached to the scales of a cone or partly surrounded by an open cup, or entirely exposed. Embryo, in most cases, having more than two cotyledons. *Gymnosperms.* **108**
Carpels closed, or mutually growing together by their edges. Seeds contained in a closed pericarp. *Angiosperms.*

Embryo without distinct cotyledons. Fleshy, leafless, parasitic · plants. Flowers mostly imperfect. *Rhizanths.* **124**

Embryo with a pair of cotyledons, rarely without any, in which case the leafless plants are parasitic, and are distinguished from the Rhizanths by having two celled ovaries, with two ovules in each cell.

Flowers having a single perianth or none. **68**
Flowers having both calyx and corolla.

Petals all cohering by their edges. *Monopetalous* or *Gamopetalous.* **3**
Petals or most of them separate or distinct. *Polypetalous, Choripetalous,* or *Dialypetalous.*

Corolla inserted on the *receptacle* or end of the flower stalk.
Thalamifloral. **45**
Corolla inserted on the calyx. **22**

3.— Corolla inserted on the calyx. **16**
Corolla inserted on the receptacle. *Corollifloral.*

COROLLIFLORAL EXOGENS.

Stamens attached to the corolla. **5**
Stamens distinct from the corolla, and attached to the receptacle.

Ovary consisting of a single carpel. Stigma surrounded by a peculiar, two-cleft covering. Ovule single. Australian herbs with radical leaves. BRUNONIACEÆ.
Ovary composed of two or more carpels.

Anthers opening by a pore at or near the top. **4**
Anthers opening by a slit lengthwise.

Plants without green leaves. MONOTROPACEÆ. p. 212.
Plants with green leaves, which in most cases, but not in all, are *punctate* with translucent dots. RUTACEÆ. p. 81.

4.—Anthers two celled, with peculiar appendages. Shrubs or low herbs with somewhat woody stems. Most of the genera have evergreen leaves. ERICACEÆ. p. 210.
Anthers one celled, without appendages. Natives of the southern hemisphere. EPACRIDACEÆ.

5.—Flowers regular. **7**
Flowers irregular.

Ovary deeply four lobed and appearing like four distinct carpels. Leaves opposite or whorled. Anther-bearing stamens, four or two. Corolla generally two lipped. Mostly aromatic herbs. LABIATÆ. p. 243.
Ovary not deeply lobed.

Fruit, a drupe or berry. Anthers two celled. Ovary consisting of two or four closely united carpels. Albumen of the seeds little or none. VERBENACEÆ. p. 241.
Fruit consisting of two achenes. Anthers one celled. Ovary two celled with one ovule in each cell. Albumen of the seeds fleshy. Leaves mostly alternate. Herbs or shrubs natives of southern Africa. SELAGINACEÆ.
Fruit a dry or fleshy capsule.

Placentas parietal. **6**
Placentas central.

Ovary one celled, with a free central placenta. Herbaceous plants, growing mostly in wet places. LENTIBULACEÆ. p. 225.
Ovary two or four celled, very rarely one celled.

Seeds with albumen. Lobes of the corolla imbricate in the bud. Flowers mostly irregular. SCROPHULARIACEÆ. p. 229.
Seeds without albumen.

Seeds without wings, partially immersed in an enlargement of the placenta. Lobes of the corolla twisted in the bud. Herbs or shrubs, not climbing. ACANTHACEÆ. p. 239.
Seeds mostly winged, numerous. Ovary surrounded by a fleshy ring. Flowers large and showy. Plants mostly climbing or twining. BIGNONIACEÆ. p. 226.

6.— Seeds winged, with no albumen. Flowers showy. Mostly climbing plants. BIGNONIACEÆ. p. 226.
Seeds not winged.

Scaly, parasitic plants without green leaves.
 OROBANCHACEÆ. p. 228.
Plants with green leaves.

Stamens five, or occasionally four. Seeds numerous. Cotyledons minute. Radicle long. Herbs or shrubs. GESNERACEÆ. p. 228.
Stamens four, two longer than the other two. Cotyledons large and fleshy. Radicle short.

Herbs. Fruit a drupe, capsule, or nut. Seeds few.
PEDALIACEÆ. p. 226.
Trees or shrubs. Fruit a berry. CRESCENTIACEÆ.

7.— Scaly plants without green leaves. Cells of the ovary ten or more with one ovule in each. Parasitic on the roots of other plants.
LENNOACEÆ.
Plants with green leaves.

Flower buds arranged in coiled cymes, which unroll from the base as the flowers open in succession. **9**
Flower buds not arranged in coiled cymes.

Ovary having three lobes or more. NOLANACEÆ. p. 266.
Ovary not lobed.

Carpels many, distinct. Tropical trees or shrubs.
ANONACEÆ. p. 43.
Carpels more than one, united. **8**
Carpel single. Parts of the flower in fours, or single.

Fruit one celled, one seeded. Ovule anatropous. Flowers perfect and complete. SALVADORACEÆ.
Fruit spuriously two celled. Flowers sometimes perfect, sometimes staminate, sometimes pistillate. Ovules amphitropous.
PLANTAGINACEÆ. p. 221.

8.—Ovary formed of four carpels or more. **13**
Ovary formed of two carpels. **10**
Ovary formed of three carpels.

Herbs. Seeds angular, oval, or winged. Sepals united. Corolla lobes twisted together in the bud. POLEMONIACEÆ. p. 260.
Shrubs. Seeds shield-like. Sepals distinct. Corolla lobes imbricate in the bud. DIAPENSIACEÆ. p. 260.

9.—Ovary having three lobes or more. BORAGINACEÆ. p. 254.
Ovary formed by two carpels. Fruit a capsule. Herbs. Albumen of seeds abundant. HYDROPHYLLACEÆ. p. 258.
Ovary formed of four carpels, with one ovule in each. Fruit a drupe. Albumen of seeds scanty. Tropical plants, mostly shrubs or trees. EHRETIACEÆ.

10.—Stamens four or more. **11**
Stamens two.

Corolla four cleft. Leaves simple. Corolla valvate in the bud.
OLEACEÆ. p. 279.
Corolla five to eight cleft, salver form, twisted or imbricate in the bud. Leaves mostly compound. JASMINACEÆ. p. 279.

11.—Leaves none. Parasitic, yellow stems, with a few minute scaly bracts and, at length, dense clusters of flowers. Embryo thread-like, coiled, acotyledonous. CUSCUTACEÆ. p. 263.
Leaves opposite, or in whorls, sometimes wanting, and then the plants are green and fleshy. **12**
Leaves alternate.

Style two or four cleft. Fruit a drupe. Seeds without albumen. Trees or shrubs. CORDIACEÆ.
Style not two cleft. Fruit a capsule, follicle, or berry.

Seeds very numerous, provided with long silky hairs, flat, contained in large follicles. Flowers in umbels. Herbs with a milky juice.
ASCLEPIADACEÆ. p. 276.
Seeds not hairy. Leaves without stipules.

Ovules few. Corolla twisted in the bud. Albumen of the seeds scanty, mucilaginous. Convolvulaceæ. p. 262.

Ovules numerous. Corolla plaited, valvate, or imbricate in the bud. Albumen of the seeds abundant, fleshy.

Corolla plaited or valvate in the bud. Solanaceæ. p. 265.
Corolla imbricate in the bud. Atropaceæ. p. 266.

12.—Anthers united to the stigma. Seeds flat, contained in a follicle, and furnished with long, silky hairs. Leaves mostly opposite.
Asclepiadaceæ. p. 276.
Anthers free from the stigma.

Stigma with an hourglass-like contraction. Mostly with a milky juice. Apocynaceæ. p. 274.
Stigma without a central contraction.

Leaves with stipules. Loganiaceæ. p. 273.
Leaves without stipules.

Corolla twisted or plaited in the bud. Leaves mostly opposite or in whorls. Seeds numerous. Gentianaceæ. p. 270.
Corolla imbricate in the bud. Hydrophyllaceæ. p. 258.

13.—Stamens alternate with the lobes of the corolla. **14**
Anther-bearing stamens opposite the lobes of the corolla.

Styles or stigmas five, rarely three or four.
Plumbaginaceæ. p. 222.
Style one.

Fruit fleshy. Trees or shrubs. Myrsinaceæ.
Fruit a capsule. Herbs. Primulaceæ. p. 222.

14.—Ovules hanging downward. **15**
Ovules erect or tending upward.

Corolla imbricate in the bud. Leaves alternate. Trees or shrubs.
Sapotaceæ. p. 220.
Corolla twisted in the bud. Convolvulaceæ. p. 262.

15.—Embryo of seeds large. Trees or shrubs with a milky juice.
SAPOTACEÆ. p. 220.
Embryo small.

Stamens as many as the divisions of the corolla.
AQUIFOLIACEÆ. p. 218.
Stamens twice as many as the divisions of the corolla.
EBENACEÆ. p. 219.

MONOPETALOUS CALYCIFLORAL EXOGENS.

16.—Calyx adherent to the ovary. **17**
Calyx free from the ovary or ovaries.

Leaves with minute stipules. Stamens five. Petals five. Ovary consisting of three or five carpels. Leaves alternate, entire. Calyx tube enlarged. Australian herbs or somewhat shrubby plants. STACKHOUSIACEÆ.
Leaves without stipules.

Carpel, one to each flower. Stigma with a cup-like covering. Leaves radical. Australian herbs. BRUNONIACEÆ.
Capsules more than one.

Carpels distinct. Leaves commonly fleshy. CRASSULACEÆ. p. 137.
Carpels combined into a single ovary.

Flowers perfect. STYRACACEÆ. p. 220.
Flowers imperfect, some staminate, some pistillate. Trees with a milky juice. PAPAYACEÆ.

17.—Ovary with three cells, but only one of them productive, there being a single ovule in the fertile cell. Fruit a single achene with the rudiments of two others. Stigmas two or three.
VALERIANACEÆ. p. 177.
Ovary consisting of two or more productive carpels. **19**
Ovary consisting of a single carpel with one ovule.

Anthers distinct. Stigma single, but sometimes two lobed. **18**
Anthers united into a tube. Flowers compound, consisting of many
florets, contained within a common, calyx-like involucrum.

Stigma two cleft. Ovule erect from the base of the carpel.
CompositÆ. p. 179.
Stigma undivided. Ovule hanging from the apex of the carpel.
CalyceraceÆ.

18.—Fruit a utricle crowned with the adherent calyx, and inclosed by
an involucel. Mostly herbs. Stamens usually four.
DipsacaceÆ. p. 178.
Fruit a berry, with an involucel. Evergreen shrubs, mostly parasitic. LoranthaceÆ. p. 292.

19.—Leaves without stipules. **20**
Leaves with stipules, or tendrils.

Stipules between opposite petioles. CinchonaceÆ. p. 173.
Stipules assuming the form of tendrils. Leaves alternate, rough.
CucurbitaceÆ. p. 158.

20.—Anthers united into a somewhat curved tube. Juice usually milky.
Stamens five. LobeliaceÆ. p. 208.
Anthers distinct.

Stamens more than two. **21**
Stamens two or one.

Stamens inserted on the tube of the corolla. Tropical trees or
shrubs. ColumelliaceÆ.
Stamens two. Filaments adhering to the style. StylidiaceÆ.

21.—Anthers opening by pores at the apex. VaccinniaceÆ. p. 211.
Anthers opening by a slit lengthwise.

Stigma surrounded by a peculiar covering or envelope.
GoodeniaceÆ.
Stigma without a covering.

Leaves in whorls of four or more. RUBIACEÆ, or GALIACEÆ. p. 173.
Leaves opposite. Corolla lobes imbricate in the bud.

CAPRIFOLIACEÆ. p. 169.

Leaves mostly alternate.

Stamens four or five. Corolla lobes valvate in the bud.

CAMPANULACEÆ. p. 209.

Stamens five or many. Corolla lobes plaited in the bud.

BELVISIACEÆ.

POLYPETALOUS CALYCIFLORAL EXOGENS.

22.— Stamens not more than twelve. **29**
Stamens more than twelve.

Calyx adherent to the ovary. **25**
Calyx wholly free from the ovary.

Leaves with stipules. , **24**
Leaves without stipules.

Carpels combined into a single pistil, often with two or more stigmas. With more than one placenta. **23**
Carpels more or less distinct, or single.

Flowers more commonly imperfect. Ovule attached to the base of
the one-celled ovary. ANACARDIACEÆ. p. 84.
Flowers perfect. Ovules either standing obliquely upward or
suspended from the apex of the cell. ROSACEÆ. p. 115.

23.— Stigmas usually two or more. Calyx commonly of two sepals
cohering at base. Leaves more or less fleshy.

PORTULACACEÆ. p. 69.

Stigma single, rarely two lobed. Calyx of more than two sepals,
tubular, remaining after the withering of the flower.

LYTHRACEÆ. p. 149.

24.— Carpels combined into a single pistil, with more than one placenta. Placentas in the axis. Stigmas commonly two or more,

PORTULACACEÆ. p. 59.

Carpels more or less distinct, or single. When there are several
they are sometimes joined together in the fruit.

Carpel single. Fruit a two valved pod. Calyx with the odd lobe on the outer side, or the side away from the main stem.

LEGUMINOSÆ. p. 94.

Carpels often many, sometimes single. Calyx with the odd lobe, when there is such, on the upper or inner side.

ROSACEÆ. p. 115.

25.— Leaves with stipules. **28.**
Leaves without stipules.

Placentæ in the axis. **26.**
Placentæ parietal.

Petals few; as many or twice as many as the lobes of the calyx, and distinct from the sepals, usually concave or hood-like.

LOASACEÆ. p. 151.

Petals numerous. Fleshy plants.

Calyx fleshy, five to sixteen parted, unlike the corolla.

MESEMBRYANTHEMACEÆ. p. 156.

Calyx lobes petal-like, numerous, with no distinct limit between sepals and petals. Plants generally leafless, or with little distinction between stems and leaves. CACTACEÆ. p. 152.

26.— Petals numerous or indefinite. Plants fleshy. Calyx five to sixteen parted, unlike the corolla. MESEMBRYANTHEMACEÆ. p. 156.
Petals few or definite.

Ovary with two cells or more. **27.**
Ovary one celled.

Fruit a one-seeded drupe. Petals linear, reflexed. Natives of India. ALANGIACEÆ.
Fruit a capsule, which is commonly one seeded. Petals ovate or rounded. Australian shrubs. CHAMÆLAUCIACEÆ

27.— Leaves with translucent dots. Style simple. Stigma undivided.

MYRTACEÆ. p. 149.

Leaves without translucent dots.

Fruit a one or few-seeded berry. Stamens numerous. Style single.
Tropical trees. BARRINGTONIACEÆ.
Fruit a many-seeded capsule. Leaves opposite. Shrubs.
PHILADELPHACEÆ. p. 134.

28.— Carpels more or less distinct, or solitary, covered by the calyx.
ROSACEÆ. p. 115.
Carpels united.

Leaves opposite. Fruit one seeded. RHIZOPHORACEÆ.
Leaves alternate. Fruit many seeded. LECYTHIDACEÆ.

29.— Calyx superior, or adherent to the ovary, or partly so. **39**
Calyx wholly inferior, or free from the ovary.

Leaves with stipules. **34**
Leaves without stipules, or with minute ones that soon drop off.

Ovary consisting of two or more carpels united into one. **30**
Ovary or ovaries simple, not composed of united carpels.

Carpels each with a scale beneath. Mostly thick-leaved plants.
CRASSULACEÆ. p. 137.
Carpels without scales beneath.

Carpels several, all perfect. CALYCANTHACEÆ. p. 130.
Carpels solitary, or all but one imperfect.

Leaves with translucent dots. AMYRIDACEÆ.
Leaves without translucent dots.

Plants with a resinous or milky juice. ANACARDIACEÆ. p. 84.
Tropical plants with watery juice. CONNARACEÆ.

30.— Placentæ in the axis. **31**
Placentæ parietal. Tropical plants.

Flowers imperfect, some staminate, some pistillate. PAPAYACEÆ.
Flowers perfect, having both stamens and pistils. TURNERACEÆ.

31.— Styles partly or wholly united. **32**
Styles distinct to the base.

Carpels each with a scale beneath. Crassulaceæ. p. 137.
Carpels two without hypogynous scales. Saxifragaceæ. p. 131.

32.— Sepals imbricate in the bud. **33**
Sepals valvate in the bud.

Stamens as many as the petals, and opposite to them.
Ruamnaceæ. p. 86.
Stamens alternate with the petals. Lythraceæ. p. 149.

33.— Sepals two. Mostly thick-leaved plants. Portulacaceæ. p. 69.
Sepals more than two.

Stigmas two to five lobed. Ovules erect or slanting upwards.
Celastraceæ. p. 87.
Stigmas minute. Ovules hanging down. Natives of southern
Africa. Bruniaceæ.

34.— Ovary consisting of two or more carpels united. **35**
Ovary or ovaries simple.

Fruit a pod. Odd sepal inferior, or on the outer side. Flowers
commonly irregular. Leguminosæ. p. 94.
Fruit not a pod. Odd sepal on the superior or inner side. Flowers regular or nearly so. Rosaceæ. p. 115.

35.— Placentæ in the axis. **36**
Placentæ parietal.

Flowers with a supernumerary fringe of abortive petals, or with
several inner fringes. Passifloraceæ. p. 157.
Flowers without a supernumerary fringe. Moringaceæ.

36.— Styles partly or wholly united. **37**
Styles distinct to the base.

Petals minute. Ovary one celled, containing a single ovule.
Illecebraceæ.
Petals conspicuous. Ovary usually with more than one cell, but if
one celled, with more than one ovule.

Leaves opposite. CUNONIACEÆ. p. 131.
Leaves alternate. SAXIFRAGACEÆ. p. 131.

37.—Sepals imbricate in the bud. **38**
Sepals valvate in the bud.

Stamens opposite the petals, and of the same number.
RHAMNACEÆ. p. 86.
Stamens alternate with the petals. Resinous trees or shrubs of tropical America. AMYRIDACEÆ.

38.—Flowers irregular. Trees or shrubs of tropical South America.
VOCHYSIACEÆ.
Flowers regular.

Leaves simple, alternate. Australian herbs. STACKHOUSIACEÆ.
Leaves compound, mostly opposite. Trees or shrubs.
STAPHYLÆACEÆ. p. 89.

39.—Leaves with stipules. **44**
Leaves without stipules.

Placentæ parietal. Fruit a juicy berry. GROSSULARIACEÆ. p. 133.
Placentæ in the axis.

Flowers not in umbels. **40**
Flowers in umbels.

Styles two. Carpels separating when ripe. Flowers arranged in flat-topped umbels. UMBELLIFERÆ. p. 162.
Styles commonly more than two. Carpels not separating when ripe. Umbels not flat topped. ARALIACEÆ. p. 166.

40.—With more than one carpel to each flower. **41**
With only one carpel to each flower.

Plants mostly parasitic. Fruit a berry, which is usually one seeded.
LORANTHACEÆ. p. 292.
Not parasitic.

Leaves with translucent dots. Fruit a capsule. Australian shrubs.
CHAMÆLAUCIACEÆ.
Leaves without translucent dots.

Petals linear, turned back. Fruit a drupe. Leaves alternate,
simple. Tropical trees or shrubs. ALANGIACEÆ.
Petals more or less rounded.

Plants with a resinous or milky juice. Ovule single.
ANACARDIACEÆ. p. 84.
Plants mostly with a watery juice. Ovules two to five.

Herbs. Stigmas four. Cotyledons short, straight, obtuse. Petals
four. HALORAGEACEÆ. p. 140.
Tropical trees or shrubs. Stigma single, undivided. Cotyledons
leaf-like, twisted or folded. COMBRETACEÆ.

41.—Calyx limb minute. HALORAGEACEÆ. p. 140.
Calyx limb conspicuous.

Carpels not separating at the apex as they ripen. **42**
Carpels separating widely at the apex.

Leaves alternate or radical. Herbs. SAXIFRAGACEÆ. p. 131.
Leaves opposite. Shrubs.

Corolla valvate in the bud. HYDRANGEACEÆ. p. 132.
Corolla infolded in the bud. PHILADELPHACEÆ. p. 134.

42.—Calyx not valvate in the bud. **43**
Calyx valvate in the bud.

Stamens opposite the petals, and of the same number.
RHAMNACEÆ. p. 86.
Stamens alternate with the petals, or not of the same number.

Fruit a capsule, nut, or berry. Corolla twisted in the bud. Albumen of the seeds none. Ovules mostly numerous.
ONAGRACEÆ. p. 141.
Fruit a drupe. Corolla valvate in the bud. Seeds albuminous.
Ovules mostly single. CORNACEÆ. p. 167.

43.— Stamens doubled downwards. Anthers elongated. Leaves ribbed. MELASTOMACEÆ. p. 148.
Stamens not doubled down. Anthers short.

Seeds very numerous, minute. ESCALLIONIACEÆ. p. 131.
Seeds one to four. BRUNIACEÆ.

44.— Placentæ parietal. Stipules becoming tendrils.
CUCURBITACEÆ. p. 158.
Placentæ in the axis.

Stamens opposite the petals, and of the same number.
RHAMNACEÆ. p. 86.
Stamens alternate with the petals, or not of the same number.

Leaves opposite. Tropical or Australian trees or shrubs.
RHIZOPHORACEÆ.
Leaves alternate. HAMAMELIDACEÆ. p. 140.

THALAMIFLORAL EXOGENS.

45.— Stamens few, or not more than twelve. **50**
Stamens numerous.

Leaves with stipules, never radical. **48**
Leaves without stipules, or else radical.

Leaves with translucent dots, and opposite or in whorls. Stamens mostly united in sets. HYPERICACEÆ. p. 61.
Leaves without translucent dots, or else alternate. Stamens distinct, or only at base cohering into a tube.

Carpels more than one, united into a single ovary. **46**
Carpels distinct, sometimes single.

Carpels immersed in a flat-topped, fleshy disc. Aquatic. Petals numerous. Flowers large. NELUMBIACEÆ. p. 46.
Carpels not immersed in the disc.

Leaves peltate, or centrally attached to the petiole. Embryo inclosed in a peculiar *vitellus* or sac. Petals and sepals three or four. Aquatic herbs. CABOMBACE.E.
Leaves not peltate. Embryo minute, not in a vitellus. Plants seldom aquatic.

Anthers opening by pores. Seeds with a peculiar attachment, or *aril*. Tropical or Australian trees or shrubs. DILLENIACEÆ.
Anthers opening lengthwise. Seeds without an aril.

Herbs or climbing plants. Flowers perfect.
RANUNCULACE.E. p. 33.
Trees or shrubs.

Flowers imperfect; some staminate, some pistillate. Leaves often with pellucid dots. SCHIZANDRACEÆ.
Flowers mostly perfect. Albumen of the seeds mottled. Sepals three. Petals six. Tropical. ANONACEÆ. p. 43.

46.— Placentæ in the axis. **47**
Placentæ parietal.

Aquatic herbs. Sepals four to six. Petals and stamens numerous. Seeds many. Embryo contained in a vitellus or sac.
NYMPHÆACEÆ. p. 46.
Not aquatic. Seeds without a vitellus.

Sepals two, rarely three; falling off soon after the flower opens. Herbs. Flowers perfect. Seeds with oily albumen.
PAPAVERACEÆ. p. 48.
Sepals usually more than two.

Seeds albuminous. Embryo straight. Trees or shrubs growing in warm countries. FLACOURTIACEÆ.
Seeds without albumen. Cotyledons curved, folded, or twisted.
CAPPARIDACEÆ. p. 56.

47.— Stigma petaloid, umbrella-like. Leaves pitcher-form. Herbs growing in wet places. SARRACENIACEÆ. p. 47.
Stigma simple, not peculiar.

Leaves compound. Trees, natives of tropical America.
RHIZOBOLACE.E.
Leaves simple.

Juice yellow, milky. Leaves opposite, leathery. Seeds without
albumen. Radicle very small. Tropical trees or shrubs.
CLUSIACE.E.
Juice not yellow.

Sepals and petals twisted in the bud. Leaves mostly opposite.
Petals five, or rarely three, ovate or round, thin, more or less
wavy or crumpled, of short duration. Seeds many, albuminous.
Herbs or shrubs. CISTACE.E. p. 60.
Sepals imbricate in the bud. Leaves mostly alternate. Trees or
shrubs.

Juice balsamic. Petioles not jointed at the base nor articulated
with the stem. Calyx persistent. Styles united into one, linear.
Stigma five lobed. Fruit a drupe. Seeds few, albuminous.
HUMIRIACE.E.
Juice watery. Petioles articulated with the stem. Sepals leathery.

Calyx deciduous, usually not green. Stigma sessile, or terminating
a short style. Seeds without albumen. Natives of tropical
America. MARCGRAVIACE.E.
Calyx usually persistent. Sepals concave. Styles two or more.
Leaves with translucent dots. TERNSTRÖMIACE.E. p. 75.

48.— Carpels distinct, commonly numerous. Trees or shrubs.
MAGNOLIACE.E. p. 42.
.Carpels united into one ovary with more than one placenta.

Calyx valvate in the bud. **49**
Calyx imbricate or twisted in the bud.

Leaves alternate, leathery. Sepals three, concave, imbricate in the
bud. Small trees or climbing shrubs, natives of Madagascar.
CHL.ENACE.E.
Leaves mostly opposite. Sepals five, twisted in the bud. Herbs
or shrubs. CISTACEÆ. p. 60.

49.—Calyx irregular, five parted, commonly enlarged in fruit. Juice resinous. Trees of tropical Asia. DIPTEROCARPACEÆ.
Calyx regular, not enlarged in fruit. Juice not resinous.

Stamens distinct. **49,** b.
Stamens partly or wholly united.

Anthers one celled. Leaves simple. Shrubs or herbs, seldom trees. MALVACEÆ. p. 70.
Anthers two celled. Tropical trees or shrubs; very rarely herbs.

Stamens arranged in a column, all perfect. Calyx twisted or imbricate in the bud. Fruit woolly or hairy within.
STERCULIACEÆ. p. 75.
Stamens rarely all anther bearing. Calyx valvate in the bud.
BYTTNERIACEÆ.

49, b.—Anthers opening lengthwise. TILIACEÆ. p. 75.
Anthers opening at the end. Petals fringed. ELÆOCARPACEÆ.

50.—Anther valves rolling upwards. BERBERIDACEÆ. p. 44.
Anthers opening lengthwise.

Leaves with stipules. **62**
Leaves without stipules.

Carpels more than one, combined into a single ovary. **52**
Carpels distinct or solitary.

Flowers all perfect. **51**
Flowers regular, some staminate, some pistillate, some perfect. Ovules mostly two or four in each ovary. Leaves commonly with translucent dots. Trees or shrubs. XANTHOXYLACEÆ. p. 82.
Flowers all imperfect.

Leaves compound. Ovules numerous. Embryo minute. Albumen of seeds copious. LARDIZABALACEÆ.
Leaves simple, entire. One ovule in each ovary. Embryo large. Albumen of seeds scanty. MENISPERMACEÆ. p. 44.

51.— Aquatic herbs, with lower leaves opposite and finely divided, immersed. Upper leaves alternate, floating, shield-like, oval or round, entire. Embryo in a membranous sac or vitellus.

CABOMBACEÆ.

Rarely growing in water. Embryo not in a vitellus.

Tropical trees or shrubs. Leaves alternate, simple, entire. Calyx three parted. Corolla six petalled. Filaments very short. Seeds with a mottled albumen. ANONACEÆ. p. 43.
Herbs, or rarely shrubs. Leaves mostly more or less divided. Seeds with a homogeneous albumen.

Sepals two. Petals four, one or two sac-like or spurred.

FUMARIACEÆ. p. 49.

Sepals more than two. Petals rarely spurred, and when so either all alike or inclosed in spurred sepals. RANUNCULACEÆ. p. 33.

52.— Aquatic. Placentæ covering the dissepiments.

NYMPHÆACEÆ. p. 46.

Not aquatic. Placenta in the axis. **54**
Placentæ parietal. Rarely growing in water.

Stamens six tetradynamous, four long and two short. Petals four, arranged cruciform. CRUCIFERÆ. p. 51.
Stamens not tetradynamous.

Disc at the base of the flower small or none. **53**
Disc large.

Stigma undivided. Fruit closed at the apex.

CAPPARIDACEÆ. p. 56

Stigma lobed. Fruit open at the apex. RESEDACEÆ. p. 57.

53.— Sepals three, soon falling off. Fruit a many-seeded capsule or pod. Stigmas three or four, sessile. PAPAVERACEÆ. p. 48.
Calyx five or six parted or leaved.

Capsule included in the persistent calyx tube. FRANKENIACEÆ.
Fruit not included in a persistent calyx.

Herbs. Leaves rolled up in the bud, with glandular, sensitive hairs.
DROSERACEÆ. p. 59.
Trees or shrubs growing in warm climates. Leaves alternate.

Ovary one celled, with several parietal placentæ. Ovules numerous. FLACOURTIACEÆ.
Ovary two or three celled, with two hanging ovules in each cell.
Anther-bearing stamens five. Abortive stamens or petals five.
Really apetalous. CHAILLETIACEÆ.

54.—Styles more or less united. **55**
Styles separate to the base.

Calyx valvate in the bud. South American herbs or shrubs.
VIVIANIACEÆ.
Calyx imbricate in the bud.

Leaves rolled up in the bud, from apex to base, furnished with
glandular, sensitive hairs. DROSERACEÆ. p. 59.
Leaves not rolled up in the bud.

Carpels attached to a *carpophore*, or prolongation of the axis beyond
the insertion of the perianth. Styles five. OXALIDACEÆ. p. 77.
Carpels attached to the receptacle.

Petals twisted in the bud. Leaves linear, sessile. Embryo straight. .
LINACEÆ. p. 77.
Petals not twisted in the bud. Leaves mostly opposite. Embryo
curved. CARYOPHYLLACEÆ. p. 63.

55.—Flowers regular or symmetrical. **57**
Flowers irregular or unsymmetrical.

Calyx not spurred. **56**
Calyx with a spur or blunt projection, not wholly green. Part of
the petals inserted on the calyx. Herbs.

Stamens five. Carpels five. BALSAMINACEÆ. p. 81.
Stamens eight. Carpels two or three. TROPÆOLACEÆ. p. 81.

56.—Ovary one celled. Fruit not winged. Shrubs. KRAMERIACEÆ.
Ovary with two cells or more.

Herbs or shrubs.   Carpels two.   Flowers very irregular, having
a papilionaceous aspect.   Filaments usually cohering into a tube.
Leaves simple, entire, opposite.   Anthers opening by pores at
the apex.   POLYGALACEÆ.   p. 92.

Trees or shrubs.   Carpels two.   Fruit a samara.   Anthers opening
lengthwise.   Disc fleshy.   Stigma two cleft.   Petals without
appendages at base.   Leaves opposite.   ACERACEÆ.   p. 91.

Trees or shrubs, rarely herbs.   Ovary formed of two to four carpels.   Petals with appendages at base.   Disc fleshy.   Leaves
mostly compound and alternate.   SAPINDACEÆ.   p. 88.

57.—Carpels attached to a *gynophore*, an enlargement or prolongation
of the axis beyond the insertion of the perianth.   **61**
Carpels inserted on the ordinary receptacle.

Calyx imbricate in the bud.   **59**
Calyx valvate in the bud.

Anthers opening by pores.   **58**
Anthers opening by slits lengthwise.

Calyx enlarging in fruit.   Stamens opposite the petals.   Fruit a
one-seeded drupe.   OLACACEÆ.
Calyx continuing small.   Stamens alternate with the petals..
ICACINACEÆ.

58.—Ovary five or six celled, with many ovules in each cell.   Leaves
evergreen.   PYROLACEÆ.   p 212.
Ovary one or two celled, with one or two ovules in each cell.   Slender Australian shrubs.   TREMANDRACEÆ.

59.—Stamens distinct.   **60**
Filaments united at base.

Styles two to four, distinct.   Seeds hairy or downy.   Leaves small,
somewhat fleshy, alternate, sessile.   TAMARICACEÆ.   p. 63.
Styles united into one.

Leaves abounding in translucent dots. Fruit a berry, usually large, orange-like. AURANTIACEÆ. p. 83.
Leaves dotted not at all, or only on the margin.

Seeds winged. Corolla twisted in the bud. Fruit a capsule. Tropical trees with dense, odoriferous, and colored wood. Stamens eight or ten. CEDRELACEÆ.
Seeds not winged.

Connective of the anthers remarkably thickened at base. Filaments united at base, free at top. Leaves simple. Fruit a drupe. Juice balsamic. Trees or shrubs of tropical America. HUMIRIACEÆ.
Connective not thickened. Filaments wide and flat, and united into a decided tube. Leaves mostly pinnate. Stamens four, fleshy at base, opposite the petals. Juice yellow, resinous. Ovules few in each cell. MELIACEÆ. p. 84.
Fruit a capsule. Flowers diœcious. Ovary four celled. Ovules numerous. CLUSIACEÆ.

60.— Seeds winged. Fruit a capsule. Corolla twisted in the bud. Stamens ten. Leaves pinnate, sometimes showing minute, transparent dots. Tropical trees, usually with colored wood. CEDRELACEÆ.
Seeds not winged.

Leaves with translucent dots. Filaments widened at base. Stigma undivided. Fruit an orange-like berry. Seeds without albumen. Trees or shrubs abounding in glandular drops of volatile oil. Natives of tropical Asia. AURANTIACEÆ. p. 83.
Leaves without translucent dots.

Flowers imperfect, diœcious. Petals five. Stamens five or ten. Ovary five celled, with many ovules in each cell. Seeds with a fleshy aril. Leaves opposite. Trees of tropical America. CLUSIACEÆ.
Flowers sometimes perfect, sometimes imperfect. Stamens three or two. Petals three or two. Ovary two to nine celled, with one ovule in each cell, ascending from the base of the cell. Shrubs. EMPETRACEÆ.
Flowers perfect.

Ovary two or four celled with a single, hanging ovule in each cell. Petals five. Stamens five or ten. Style short or none. Leaves alternate. North American shrubs or trees. CYRILLACE.E.
Ovary two to five celled with more than one ovule in each cell.

Leaves provided with glandular, sensitive hairs, and rolled up in the bud, or else smooth but furnished with marginal spines, and slowly folding together when touched. DROSERACE.E. p. 59.
Leaves not sensitive nor provided with glandular hairs.

Calyx persistent. Ovary five celled. Style very short. Fruit a drupe. Trees of Madagascar. BREXIACE.E.
Calyx deciduous. Ovary generally two celled. Fruit a capsule or berry. PITTOSPORACE.E. p. 57.

61.—Stamens arising from scales. Leaves alternate, usually pinnate. Styles united at the apex, often twisted. Fruit consisting of drupes. Tropical trees or shrubs, distinguished by the bitterness of all parts. SIMARUBACE.E. p. 83.
Stamens not arising from scales.

Flowers imperfect in part, some staminate, some pistillate, rarely some perfect. Leaves usually with translucent dots. Calyx valvate or imbricate in the bud. Ovaries three to five, with two or rarely four ovules in each cell. Tropical trees or shrubs.
XANTHOXYLACE.E. p. 82.
Flowers all perfect. Mostly herbs.

Calyx valvate in the bud. Leaves pinnate or bipinnate, without translucent dots. Style single. Stigma three to five cleft. Fruit consisting of three to five achenia. LIMNANTHACE.E. p. 77.
Calyx imbricate in the bud.

Styles united at the apex. Leaves mostly with translucent dots. Anthers with a minute gland at the apex. Calyx persistent.
RUTACE.E. p. 81.
Styles united only at the very base. Leaves mostly compound Styles five, persistent. Juice acid. OXALIDACE.E. p. 77.

62.—Placentæ in the axis. **63**
Placentæ parietal.

Carpels single. Fruit a berry.

BERBERIDACEÆ. p. 44.

Carpels three, united. Flowers mostly irregular. Leaf edges often rolled inward in the bud. VIOLACEÆ. p. 58.

63.— Carpels borne on a gynophore, or prolongation of the axis, beyond the insertion of the perianth. **67**
Carpels not borne on a gynophore.

Calyx valvate in the bud. · **65**
Calyx imbricate in the bud.

Trees or shrubs. **64**
Herbs.

Petals inconspicuous. Marsh plants. ELATINACEÆ. p. 63.
Petals conspicuous. Stipules dry and scarious.

ILLECEBRACEÆ. p. 64.

64.— Ovary four to ten celled. Leaves compound, mostly opposite. Stamens distinct. ZYGOPHYLLACEÆ.
Ovary five celled. Stamens united. Leaves alternate, simple.

STERCULIACEÆ. p. 75.

Ovaries distinct, or ovary two or three celled.

Flowers in heads, with a common involucrum. Leaves alternate, simple. Stamens ten. Shrubs of Madagascar. CHLÆNACEÆ.
Flowers without a common involucrum.

Stamens three. Filaments flat, widened at base. Petals five. Sepals five. Flowers with an expanded disc between the petals and the ovary. Leaves opposite. Tropical trees or shrubs.

HIPPOCRATEACEÆ.

Stamens more than three.

Calyx usually glandular. Petals with long claws. Filaments commonly united. Carpels usually three. Petals five. Sepals five.

MALPIGHIACEÆ.

Calyx not glandular.

Calyx densely covered with stellate hairs.   Stamens united at
base. STERCULIACEÆ. p. 75.
Calyx not hairy.

Calyx mostly irregular.   Stamens usually eight.   Filaments slender.
Petals inserted on a fleshy disc.   Embryo curved.

                    SAPINDACEÆ.   p. 88.

Flowers regular, without a fleshy disc.   Corolla with scales between
the petals. Stamens ten.   Filaments flattened at base and united
into a tube. Fruit a drupe. Embryo straight.

                    ERYTHROXYLACEÆ.

65. — Stamens united at base.   **66**
     Stamens distinct.

Stamens opposite the petals.   Stamens four or five.   Fruit a berry.
Mostly climbing shrubs with tendrils opposite the leaves.

                    VITACEÆ.   p. 85.
Stamens alternate with the petals.   Leaves alternate.   Not climbing.

                    TILIACEÆ.   p. 75.

66. — Plants not hairy.   Calyx persistent.   Mostly climbing shrubs.
     Fruit a juicy berry.   VITACEÆ.   p. 85.
With parts of the plant hairy or downy, very rarely smooth.   Fruit
usually a capsule.

Anthers distinctly two celled, with the cells straight, parallel, and
opening lengthwise or by an orifice near the apex. BYTTNERIACEÆ.
Anthers curved, or with the cells divergent. STERCULIACEÆ. p. 75.

67. — Leaves compound, opposite.   ZYGOPHYLLACEÆ.
'   Leaves simple.

Fruit a drupe or berry.   Tropical trees or shrubs.   OCHNACEÆ.
Fruit consisting of five beaked capsules; when ripe, separating at
base, but remaining with the permanent styles adherent to the
apex of the carpophore.   Mostly herbs.   GERANIACEÆ.   p. 77.

APETALOUS EXOGENS.

68.—Flowers not in aments. **70**
Flowers in aments. Trees or shrubs.

Leaves opposite, simple. Staminate and pistillate flowers on different plants; both kinds in aments. Stamens four. Ovary, one celled, with two ovules. Fruit a berry, crowned with the two persistent styles. California shrubs. GARRYACEÆ. p. 167.
Leaves alternate.

Leaves pinnate. Stipules none. Staminate flowers in aments. Calyx adherent to the ovary. JUGLANDACEÆ. p. 300.
Leaves simple.

Staminate flowers with a single stamen. Leaves evergreen. Perianth very small, four parted. Stigmas two or three. ·Tropical trees or shrubs. LACISTEMACEÆ.
Stamens two or more.

Style wanting. **69**
Style present, though sometimes very short.

Ovary adherent to the scale-like perianth, with two cells or more, but becoming at length a one-celled, one-seeded fruit, which is partly or wholly inclosed in the cup-like or leaf-like involucrum. Style undivided. Stigmas two or more. Monœcious.
CUPULIFERÆ. p. 302.
Perianth none.

Style single, very short, with two long, spreading, rough stigmas. Ovule single, erect, orthotropous. ·Commonly monœcious. Leaves with resinous dots, aromatic. MYRICACEÆ. p. 305.
Styles two or more.

Styles distinct, persistent. Ovary two celled. Ovules numerous. Leaves with stipules of short duration. Seeds peltate, winged. Monœcious. ALTINGIACEÆ.
Styles more or less united. Stigmas two or three lobed. Seeds numerous, hairy or downy. Diœcious. SALICACEÆ. p. 307.

69.—Monœcious. Fertile flowers in cone-like spikes. Stamens four. Stigmas two, long, thread-like, spreading. Fruit consisting of one-celled, one-seeded nutlets attached to the bracts of the cone. Stipules falling off early. BETULACEÆ. p. 306.
Diœcious.

Flowers with a three to five-parted perianth. Stamens opposite the lobes of the perianth. Ovary surrounded at base by a glandular ring, one celled, containing two ovules. Stigma flat, three to five lobed. Fruit a one-seeded, berry-like drupe, crowned with the persistent stigma. STILAGINACEÆ.
Perianth four to six leaved, imbricate in the bud. Stamens two to five. Stigmas two to four. Ovary two celled, with two ovules in each cell. SCEPACEÆ.

70.—Having a perianth. **73**
Having no proper calyx or corolla.

Carpels two or more. **71**
Carpel single.

With a style. Stem jointed. Leaves sessile, in whorls, much divided. Monœcious. Aquatic. CERATOPHYLLACEÆ.
Style none.

Ovule erect. Embryo in a vitellus or sac. Fruit a berry.
PIPERACEÆ.
Ovule hanging. Embryo not in a vitellus. Fruit a drupe.
CHLORANTHACEÆ.

71.—Growing in the water or in wet places. **72**
Growing on dry land.

Ovaries numerous, one celled. Stamens many, densely crowded. Stipules none. Monœcious trees. PLATANACEÆ. p. 300.
Ovary two or three celled, with one ovule in each cell. Seeds without wings. EUPHORBIACEÆ. p. 293.

72.—Leaves opposite, sessile, upper and lower ones often different. Stipules none. Styles two. Ovary four celled, with one ovule in each cell. CALLITRICHACEÆ.

Leaves alternate or all radical.

Leaves petiolate. Ovary consisting of three to five carpels, partly coherent. Ovules orthotropous. Embryo obcordate, contained in a vitellus or sac. Flowers perfect. SAURURACE.E. p. 293. Leaves decurrent. Ovary two or three celled. Ovules many, amphitropous, very minute. Styles persistent. PODOSTEMONACE.E.

73.—Ovary partly or wholly adherent to the calyx, or inferior. **99** Ovary or ovaries wholly free from the calyx, or superior.

Leaves with stipules. **92** Leaves without stipules.

Flowers imperfect, some staminate, some pistillate. **85** Flowers perfect.

Carpels distinct or single. **80** Carpels more than one, united into a single ovary.

Growing in water. Flowers contained in a spatha. Carpels two or three. Styles as many as the carpels, distinct, persistent. Ovules numerous, very minute. PODOSTEMONACE.E. Not aquatic.

Anthers opening by valves which, roll outwards. Flowers with a fleshy disc at the bottom of the calyx. Ovary single, with but one ovule, though formed of three carpels united by their edges. LAURACE.E. p. 290. Anthers opening lengthwise.

Sepals two, distinct, colored, falling off soon after opening. Stamens eight or more. Stigma two lobed. Juice milky. PAPAVERACE.E. p. 48. Sepals more than two. Juice watery.

Stamens few. **75** Stamens numerous.

Placentæ parietal. **74** Placentæ central or basal.

One ovule in each cell. Styles two or more.

PHYTOLACCACEÆ.   p. 584.

Many ovules in each cell.   Plants with fleshy leaves.

PORTULACACEÆ.   p. 69.

74.—Style present.   Stigma angular or cleft.   BIXACEÆ.
Style none or very short.

Stigma round, head-like.   CAPPARIDACEÆ.   p. 56.
Stigma radially three to ten parted.   Seeds six to twenty.

FLACOURTIACEÆ.

75.—Ovary with two or more ovules in each cell.   **77**
Ovary with one ovule in each cell.

Style one or none.   **76**
Styles three or more.

Ovules attached to the base of the carpel.   Herbs.

PHYTOLACCACEÆ.   p. 284.

Ovule hanging from the apex of the carpel.   Shrubs or herbs.

PORTULACACEÆ.   p. 69.

76.—Calyx valvate in the bud.   Stigma undivided.   Ovary surrounded
by a fleshy disc.   RHAMNACEÆ.   p. 86.
Calyx imbricate in the bud.

Thick-leaved plants.   Leaves opposite or in whorls.

PORTULACACEÆ.   p. 69.

Thin-leaved plants.   Leaves alternate.   Stamens usually more than
the divisions of the calyx.   Sepals commonly unequal.

SAPINDACEÆ.   p. 88.

77.—Style one or none.   **78**
Styles or stigmas two or more.

Styles four.   Carpels four with two ovules in each.   Perianth colored, four cleft, swelled out at base, valvate in the bud, persistent.   Evergreen shrubs of southern Africa.   PENÆACEÆ.
Herbs or shrubs with fleshy leaves.   Ovary with many ovules in
each cell.   PORTULACACEÆ.   p. 69.

78.—Calyx imbricate in the bud. **79**
Calyx valvate in the bud.

Stamens five, inserted on the receptacle. Anthers turned outwards. Calyx soon withering. Leaves alternate. Australian shrubs.
BYTTNERIACEÆ.
Stamens more or less than five, inserted on the calyx tube. Anthers turned inwards. Calyx tubular, persistent.
LYTHRACEÆ. p. 149.

79.—Thick-leaved plants. Albumen of seeds mealy or fleshy.
PORTULACACEÆ. p. 69.
Thin-leaved plants. Seeds without Albumen.

Stamens eight. Stigmas three or single. Calyx five cleft.
SAPINDACEÆ. p. 88.
Stamens five or ten. Stigma single. Ovary one celled, with two parietal placentæ. Ovules two. Calyx long, tubular, or urceolate, colored. AQUILARIACEÆ.
Stamens ten. Stigmas five. Ovary five celled, with two ovules in each cell. GERANIACEÆ. p. 77.

80.—Style and stigma single. **82**
Stigmas two or more. Styles sometimes distinct, sometimes united into one.

Stamens inserted on the receptacle, or on the base of the perianth, or on a disc. **81**
Stamens inserted on the tube of the perianth.

Stamens in the throat of the perianth. Perianth tubular, funnelform, or urceolate, persistent, its base becoming hardened around the fruit. Leaves opposite, growing together at base.
SCLERANTHACEÆ. p. 64.
Stamens inserted on the middle of the perianth tube, and opposite its divisions. Perianth not hardened in the fruit. Leaves alternate. BASELLACEÆ.

81.—Flowers in heads, the outer ones sterile, the inner ones perfect. Calyx five cleft. Stamens ten. Leaves opposite. Herbs.
CARYOPHYLLACE.E. p. 63.
Flowers not in heads made up of unlike kinds.

Stem jointed, the nodes being usually covered by *ochreæ*, or colored membranous sheaths. Leaves rarely opposite. Styles commonly three. Ovule orthotropous. POLYGONACE.E. p. 287.
Stem not jointed, or, if so, without ochreous sheaths at the nodes

Stamens opposite the divisions of the calyx, or else fewer in number. Ovule amphitropous. CHENOPODIACE.E. p. 284.
Stamens alternate with the divisions of the calyx, or more numerous. Ovule campylotropous. PHYTOLACCACE.E. p. 284.
Stamens many. Ovule anatropous. RANUNCULACEÆ. p. 33.

82.—Leaves alternate, entire, covered with dry, shield-like scales. Flowers regular. Perianth tubular at base, becoming fleshy or woody in the fruit. Trees or shrubs. EL.EAGNACE.E. p. 292.
Leaves not scaly.

Perianth tubular. **83**
Perianth not tubular.

Perianth dry, mostly five leaved, persistent. Flowers with two or three bracts at base. Seeds flattened, kidney-shaped, with mealy albumen. AMARANTHACE.E. p. 286.
Perianth not dry and persistent.

Stamens four, inserted on the calyx. opposite its divisions. Sepals four, distinct or slightly cohering at base. valvate in the bud. Ovule anatropous. PROTEACE.E.
Stamens five, united at base. Calyx five parted, globose, ovoid, colored within and on the margins of the lobes. Leaves in whorls. Ovules amphitropous. CARYOPHYLLACE.E. p. 63.

83.—Stamens inserted on the receptacle. **84**
Stamens inserted on the calyx tube.

Stamens five. Calyx five parted, tube funnel-form, imbricate in the bud, persistent. Scleranthace.e. p. 64.

Stamens four. Calyx four cleft, or two lipped, valvate in the bud.
Proteace.e.

Stamens four, eight, or more. Calyx imbricate in the bud, often with a fleshy disc in its base, and with petaloid scales in the throat, colored. Thymeleace.e. p. 291.

84.— Perianth dry, five cleft. Stamens cohering to a tube. Stigma sessile. Seeds kidney-shaped. Amaranthace.e. p. 286.

Perianth corolla-like, tubular, funnel-form, or salver-form, plaited in the bud. Style present. Nyctaginace.e. p. 283.

85.— Leaves covered with dry, shield-like scales. Perianth becoming fleshy or hard in the fruit. El.eagnace.e. p. 292.

Leaves terminating in a cirrhus which is commonly enlarged to a pitcher-like extension. Stamens cohering to a tube. Anthers sixteen. Ovary four celled. Stigma sessile. Tropical shrubs.
Nepentiiace.e.

Leaves not scaly nor furnished with pitcher-bearing cirrhi.

Flowers regular. **86**
Flowers somewhat irregular or unsymmetrical.

Carpels three or more. Calyx five cleft. Fruit a capsule, one seeded or with one seed in each cell. Leaves alternate.
Sapindace.e. p. 88.

Carpels two. Fruit a samara. Stigma two cleft. Leaves opposite.
Acerace.e. p. 91.

86.— Anthers opening by a slit lengthwise, or rarely by a pore. **87**
Anthers opening by valves which roll upwards.

Ovary one celled, but formed of three carpels united by their edges, with one ovule suspended from the apex. Leaves alternate, entire. Laurace.e. p. 290.

Ovaries many, distinct, one celled, with one erect ovule in each. Leaves opposite, serrate. Atherospermace.e.

87.—Ovary single, or consisting of united carpels. **89**
Ovaries two or more distinct.

Ovaries four or more with one ovule in each. **88**
Ovaries three. Calyx of six distinct sepals.

With one ovule in each ovary. Diœcious. Leaves simple, alternate, entire. MENISPERMACEÆ. p. 44.
With many ovules in each ovary. Monœcious. Leaves compound.
Stamens six. Stigmas sessile. Shrubs. LARDIZABALACEÆ.

88.—Stamens less than twenty. Perianth tubular. Fruit consisting of drupes or berries. Leaves evergreen. Trees or shrubs.
MONIMIACEÆ.
Stamens numerous. Calyx of four or five distinct sepals which fall off early. Fruit consisting of four or five achenia.
RANUNCULACEÆ. p. 33.
89.—Perianth withering after the fertilization of the ovary. **90**
Perianth presistent.

Anther-bearing stamens five or less. Ovary single, one celled, with one ovule. AMARANTHACEÆ. p. 286.
Anther-bearing stamens six to ten. Ovary single, one celled, with one ovule. Tropical trees or shrubs. NYCTAGINACEÆ. p 283.
Stamens numerous. Ovary one celled, with five to ten parietal placentæ and six to twenty ovules. FLACOURTIACEÆ.

90.—Ovary one celled. **91**
Ovary with two cells or more.

Plants with a milky juice. Often with thick leaves or leafless.
EUPHORBIACEÆ. p. 293.
Plants with a watery juice.

Ovary three celled with two ovules in each cell. Stigmas three, sessile. Leaves alternate, trifoliate. Trees with very hard wood.
XANTHOXYLACEÆ. p. 82.
Ovary with one ovule in each cell. Leaves simple, entire.

Calyx of three or two sepals, imbricate in the bud. Stamens two or three. Style short. EMPETRACEÆ.
Calyx four or five cleft, valvate in the bud. Stamens four or five. RHAMNACEÆ. p. 86.

91.— Stem jointed. Stamen one. Perianth two or three leaved. Stigmas two. Trees or shrubs without true leaves, but with numerous leaf-like branches or phyllodes. CASUARINACEÆ.
Stem jointed. Stamens several. Styles two or more. Ovule single. Leaves mostly alternate. POLYGONACEÆ. p. 287.
Stem not jointed.

Stamens numerous. Ovules many. BIXACEÆ.
Stamens less than twenty.

Stamens united. Diœcious. Lobes of the perianth valvate in the bud. Ovule single. Seeds with a conspicuous, fleshy aril.
MYRISTICACEÆ.
Stamens distinct.

Ovules many. Placentæ parietal. Seeds with a fleshy aril. Flowers regular, showy. PASSIFLORACEÆ. p. 157.
Ovule single, or rarely more than one.

Styles two to five. Stamens five to eight. Ovule campylotropous. Herbs of tropical America. PHYTOLACCACEÆ.
Style single. Stamens four or five.

Perianth tubular, colored. Stamens inserted in the tube or throat of the perianth. Ovule anatropous. Leaves opposite, simple.
THYMELEACEÆ. p. 291.
Perianth four or five parted or five leaved, green. Stamens inserted in the base of the perianth. Ovule anthotropous.
URTICACEÆ. p. 296.

92.— Flowers perfect, having both stamens and pistils. **93**
Flowers imperfect.

Ovaries four or five, distinct, with two ovules in each. Calyx four or five parted. Stamens eight or ten. Leaves opposite or in whorls. Trees. XANTHOXYLACEÆ. p. 82.

Ovaries two or three, distinct, with one ovule in each. Stamens numerous. Leaves alternate, compound. Rosace.e. p. 115. Carpels united or single.

Ovary with two cells or more, with one or two ovules in each cell. Albumen of seeds fleshy. Plants with a milky juice.

Euphorbiaceæ p. 293.

Ovary with two or three cells, with two hanging ovules in each cell. Seeds without albumen. Anther-bearing stamens five. With a whorl of five abortive stamens. Calyx five leaved or five toothed, colored. Tropical plants. Chailletiaceæ. Ovary one celled.

Ovules many. Placentæ parietal. Bixaceæ. Ovule single.

Styles three or more. Rosaceæ p. 115. Styles two or stigma two cleft. Stamens four or five.

Ulmaceæ. p. 296.

Style single.

Stamens ten, inserted on the receptacle. Ovule pendulous. Flowers small. Byttneriaceæ.

Stamens two to four, inserted on the calyx opposite to its lobes. Trees or shrubs with a milky juice. Seeds without albumen.

Artocarpaceæ.

Stamens four or five, inserted near the base of the calyx. Albumen of seeds copious, fleshy. Juice commonly watery.

Urticaceæ. p. 296.

93 — Ovary with more than one cell, or ovaries more than one. **96**
Ovary one celled, with more than one ovule. **94**
Ovary one celled, with one ovule.

Stipules dry and colored, sheathing the stem at the nodes.

Polygonaceæ. p. 287.

Stipules not dry and sheathing.

Leaves pinnate. Stamens inserted on the calyx. Calyx tubular. Ovule pendulous. Rosaceæ. p. 115.
Leaves simple or very rarely ternate or digitate.

Leaves opposite. Seeds contained in a utricle. Stamens four, opposite the lobes of the calyx. Divisions of the calyx not all alike. Style short, two cleft. CARYOPHYLLACEÆ. p. 63.
Leaves alternate. Seeds not contained in a utricle.

Leaves entire. Stamens four, eight, or numerous. Style short or none. Fruit an achene or berry. PETIVERIACEÆ.
Leaves mostly serrate. Styles two or sessile stigma two cleft. Stamens four or five. Fruit a drupe or a capsule.
ULMACEÆ. p. 296.

94.— Leaves simple. **95**
Leaves compound. Stamens inserted on the calyx.

Stamens three to ten. Style one. Stigma one. Fruit a pod with parietal placentæ. LEGUMINOSÆ. p. 94.
Stamens five. Styles two. Leaves opposite. Seeds very small. Trees. CUNONIACEÆ. p. 131.

95.— Stamen one. Herbs with opposite leaves and solitary flowers in the axils. CARYOPHYLLACEÆ. p. 63.
Stamens more than one.

Leaves fleshy. Stamens seven to nine. Calyx tubular, five parted.
PORTULACACEÆ. p. 69.
Leaves with translucent dots. Stamens eight to thirty, inserted in the top of the tubular calyx. Calyx persistent. SAMYDACEÆ.
Leaves neither fleshy nor dotted.

Stamens numerous, inserted on the receptacle. Seeds with an aril. Fruit a capsule. PASSIFLORACEÆ. p. 157.
Stamens inserted on an enlarged or fleshy receptacle. Fruit a berry.

Albumen of seeds none. Calyx four leaved or four cleft, valvate in the bud. Seeds kidney-shaped. CAPPARIDACEÆ. p. 56.
Albumen of seeds abundant. Calyx three to seven leaved, or four to ten cleft. Sepals imbricate in the bud. BIXACEÆ.

96.— Stamens united into a tube. Calyx five parted, with two bracts at base. Ovary five celled. Ovules many. Placentæ central. Style one, stigma one. Sterculiace.e. p. 75.
Stamens distinct or nearly so.

Leaves rigid or fleshy. Ovary five celled, with many ovules in each cell. Stigmas three to five. Stamens twelve or more, inserted on the calyx. Shrubs. Portulacace.e. p. 69.
Leaves not fleshy.

Leaves simple. **97**
Leaves compound.

Ovary two celled, with many ovules in each cell. Styles two. Stamens ten or more. Leaves opposite. Trees. Cunoniac.e. p. 131.
Ovaries distinct, one celled with one ovule in each. Calyx tubular, narrowed in the throat. Stamens one to five. Leaves alternate. Herbs or shrubs. Rosace.e. p. 115.

97.— Stamens more than ten, inserted on an enlarged disc. Ovary four celled, with several ovules in each cell. Placentæ central. Styles one to four. Tropical trees. Sapindace.e. p. 88.
Stamens not more than ten.

Ovaries five, with three ovules in each. Stamens ten. Australian shrubs. Byttneriace.e.
Ovary single.

Ovary with one ovule in each cell. **98**
Ovary with two or more ovules in each cell.

Stamens five, all bearing anthers. Ovary three celled. Calyx petal-like. Seeds with a fleshy strophiole. Australian shrubs.
Byttneriace.e.
Stamens five, inserted on the calyx, with five stamens not anther bearing. Ovary two or three celled with two pendulous ovules in each cell. Calyx five leaved or five toothed, colored within.
Chailletiace.e.

98.— Ovary two celled. Styles two. Fruit a winged samara. Trees or shrubs. ULMACEÆ. p. 296.
Ovary three celled. Stamens four or five. RHAMNACEÆ. p. 86.

99.— Leaves without stipules. **101**
Leaves with stipules.

Flowers perfect, having both stamens and pistils. **100**
Flowers imperfect.

Trees or shrubs with a milky juice. Ovary single, with one ovule. Stamens few. ARTOCARPACEÆ.
Herbs with a watery juice. Ovary three celled. Styles three. Seeds many. Leaves mostly unsymmetrical. BEGONIACEÆ. p. 161.

100.— Ovary one celled. Placentæ parietal. Styles two to five. Stamens inserted in the top of the calyx. Calyx ten to thirty parted. Trees or shrubs. HOMALIACEÆ.
Ovary three to six celled. Ovules many. Styles one to four. Stamens six to twelve. Herbs. ARISTOLOCHIACEÆ. p. 282.
Ovary two celled.

Ovary with one ovule in each cell. Styles one to three. Stamens five. Calyx five cleft. Shrubs. RHAMNACEÆ. p. 86.
Ovary with several ovules in each cell. Styles two. Stamens five to ten. Leaves opposite. Shrubs or trees. CUNONIACEÆ. p. 131.

101.— Flowers perfect. **105**
Flowers imperfect.

Ovary adherent to calyx, one celled. **102**
Ovary free from calyx.

Ovary one celled. Ovules few. Calyx five parted. Herbs growing near the sea or on salt plains. PRIMULACEÆ. p. 222.
Ovary two celled, with two ovules in each cell. Stigma two cleft. Flowers diœcious. Trees. RHAMNACEÆ. p. 86.
Ovary one to four celled. Stigma usually three or four lobed. Stamens three to six. Shrubs or small trees.
SANTALACEÆ. p. 292.

102.—Stamen one, inserted on the calyx. Fruit a one-seeded, fleshy drupe, crowned with the persistent calyx.

HALORAGEACEÆ. p 140.

Stamens more than one.

Style and stigma one. **103**
Styles more than one.

Ovary one celled. Ovules many. Styles three to five. Placentæ parietal. Stamens four to fifteen. DATISCACEÆ. Ovary two to six celled. Stamens four, inserted on a glandular receptacle. ARALIACEÆ. p. 166.

103.—Ovules two or more. **104**
Ovules single.

Albumen of seeds none. Fruit a drupe. Leaves alternate. Trees or shrubs. ILLIGERACEÆ. Albumen of seeds abundant. Fruit a berry. LORANTHACEÆ. p. 292.

104.—Albumen of seeds none. Stamens eight or ten. Calyx valvate in the bud. Fruit a drupe. COMBRETACEÆ. Albumen of seeds fleshy. Fruit one seeded. SANTALACEÆ. p. 292.

105.—Style single. **106**
Styles or sessile stigmas two or more.

Leaves fleshy. Fruit a drupe or nut. TETRAGONIACEÆ. p. 156. Leaves somewhat fleshy. Fruit a capsule. PORTULACACEÆ. p. 69. Leaves not fleshy.

Ovary one celled. Ovules two to five. Leaves alternate. Shrubs or small trees. HOMALIACEÆ. Ovary one or two celled. Ovules many. Styles two.

Shrubs. Leaves simple, opposite. Stamens eight or ten.

CUNONIACEÆ. p. 131.

Herbs. Stamens eight to fourteen. SAXIFRAGACEÆ. p. 131.

106.— Aquatic herbs. Ovary one to four celled, with one ovule in each cell. Stamens one to four. Leaves opposite or in whorls.
HALORAGEACEÆ. p. 140.
Not aquatic, and mostly shrubs or trees.

Ovary with three cells or more. **107**
Ovary one celled.

Calyx free from the ovary. Ovules four. Berry four seeded. Stamens many. Leaves alternate. Trees. HOMALIACEÆ.
Calyx adherent to the ovary.

Ovule single. Stamens five. Filaments with a gland at base.
ILLIGERACEÆ.
Ovules many. Capsule one celled, one seeded. Stamens fifteen to eighteen. Shrubs. HOMALIACEÆ.
Ovules two to five.

Seeds without albumen. COMBRETACEÆ.
Albumen of seeds fleshy, abundant. SANTALACEÆ. p. 292.

107.— Ovary four celled. Stamens numerous. Style four cleft. Trees with alternate leaves. MYRTACEÆ. p. 149.
Ovary three to six celled, usually with many ovules in each cell. Stigmas mostly six cleft. Herbs. ARISTOLOCHIACEÆ. p. 282.
Ovary three celled, with one ovule in each cell. Stamens five. Style three cleft.

Leaves very small. Capsule included in the calyx.
TETRAGONIACEÆ
Leaves of ordinary size, opposite. RHAMNACEÆ. p. 86.

GYMNOSPERMS.

108.— Stem jointed. Leaves opposite. GNETACEÆ.
Stem none. Leaves pinnate. CYCADACEÆ. p. 309.
Stem not jointed.

Ovules attached to scales or bracts which are usually numerous and spirally arranged in imbricated cones; but sometimes valvate and jointly developed into a somewhat fleshy, drupe-like fruit. Leaves mostly narrow, sometimes closely imbricated.

CONIFERÆ.   p. 309.

Ovule single, erect, partially immersed in a berry-like cup.

TAXACEÆ.   p. 310.

AGLUMACEOUS ENDOGENS.

109.—Flowers not on a spadix.   **111**
Flowers on a spadix.

Herbs.   **110**
Shrubs or trees.

Flowers having both calyx and corolla.   Ovary formed of three carpels usually.   Pollen elliptical.   PALMÆ.   p. 316.
Flowers having no corolla.   Style none.   Pollen globular.

PANDANACEÆ.

110.— Flowers perfect, and having a perianth.   ORONTIACEÆ.   p. 318.
Flowers perfect, with no perianth.   NAIADACEÆ.   p. 316.
Flowers imperfect, with no calyx nor corolla.

Fruit dry.   Filaments long, with wedge-shaped anthers.   Scales or hairs in place of a perianth.   TYPHACEÆ.   p. 318.
Fruit juicy, berry-like.

Spadix not adherent to the spatha.   ARACEÆ.   p. 317.
Spatha adherent to the spadix.   Seeds with a thick-ribbed covering.   PISTIACEÆ.   p. 316.

111.—Proper perianth none.   Flowers imperfect.   Aquatic.

NAIADACEÆ.   p. 316.

Perianth adherent to the ovary.   **117**
Perianth free from the ovary.

Leaves parallel veined.   **113**
Leaves apparently net-veined.

Ovary one celled. Exotic shrubs. **112**
Ovary three to five celled.

Flowers perfect. Perianth with six or eight divisions, only half of
them petal-like. Leaves not articulated with the stem.
TRILLIACEÆ. p. 337.
Flowers imperfect. Perianth with six divisions, all petal-like.
Leaves articulated with the stem. SMILACEÆ. p. 336.

112.—Parts of the flower in twos or fours. Placenta at the base of the
ovary. ROXBURGHIACEÆ.
Parts of the flower in threes. Placentæ parietal. PHILESIACEÆ.

113.—Outer perianth petal-like. **115**
Outer perianth green, or sometimes glume-like.

Carpels more or less distinct. **114**
Carpels united into a single ovary.

Perianth consisting of three whorls, each of three parts, usually dry
and glume-like. Plants resembling the grasses.
JUNCACEÆ. p. 349.
Perianth consisting of unlike whorls.

Anthers one celled. Placentæ parietal. MAYACACEÆ.
Anthers two celled. Placentæ in the axis.

Stigma single. Filaments usually hairy. Stem round, with nodes.
COMMELYNACEÆ. p. 350.
Stigmas two or more.

Flowers perfect. Stamens six. Stigmas three. Seeds numerous.
Leaves stiff, channeled, often scurfy. BROMELIACEÆ. p. 329.
Flowers imperfect, very small, in heads. Stamens two to six.
Stigmas two to six. Seeds solitary. Leaves linear, fleshy.
ERIOCAULONACEÆ. p. 352.

114.—Placentæ parietal, forming a network on the inner surface of the
ovary. Fruit many seeded. Stamens nine or many. Ovaries
six or many. Herbs. BUTOMACEÆ. p. 320.
Placentæ in the axis, or at the base of the ovary. Ovules solitary
or in pairs.

Anthers opening on the outer side. Embryo of the seeds straight, with a slit on the side near the base. JUNCAGINEÆ. p. 319.
Anthers opening on the inner side. Embryo curved.
ALISMACEÆ. p. 319.

115.—Carpels combined into a single ovary. **116**
Carpels more or less distinct.

Anthers opening on the outer side. MELANTHACEÆ. p. 337.
Anthers opening on the inner side.

Perianth of two parts. PHILYDRACEÆ.
Perianth of six parts.

Stamens nine or more. Ovaries six, with many ovules in each.
Flowers perfect. BUTOMACEÆ. p. 320.
Stamens six. Flowers imperfect, monœcious. Ovaries numerous, on a rounded receptacle, with one ovule in each.
TRIURIDACEÆ.

116.—Anthers opening on the outer side. Outer perianth glume-like.
Plants resembling the grasses. XYRIDACEÆ. p. 351.
Anthers opening on the inner side.

Flowers more or less irregular. Perianth rolling inwards after opening. Aquatic. PONTEDERIACEÆ. p. 322.
Flowers regular. Perianth not rolling inwards.

Inner perianth minute, of a single lobe, or urn-shaped, or five toothed. GILLIESIACEÆ.
Inner and outer perianth alike, or nearly so.

Perianth tubular, commonly rough or hairy outside.
HÆMODORACEÆ.
Perianth smooth inside and outside. Anthers versatile.
LILIACEÆ. p. 337.

117.—Leaves net-veined. Flowers regular, inconspicuous, imperfect.
Staminate flowers with six stamens. Fertile flowers with three carpels. Styles three, nearly or wholly distinct.
DIOSCORACEÆ. p. 335.
Leaves mostly parallel-veined.

Stamens distinct from the pistil. **118**
Stamens attached to the pistil or *gynandrous.*

Ovary one celled. Placentæ parietal. Flowers very irregular,
mostly showy. Pollen grains adhering in masses called *pollinia.*
ORCHIDACEÆ. p. 323.
Ovary three celled. Placentæ in the axis. Pollen grains distinct.
Flowers inconspicuous, arranged in many-flowered, nodding
racemes. APOSTASIACEÆ.

118.—Veins of the leaves parallel with the midrib. **120**
Veins of the leaves running obliquely from the midrib to the
margin.

Flowers perfect, with a single anther-bearing stamen. **119**
Stamens three or more.

Flowers irregular, perfect, showy. Stamens six. Ovary three celled.
Placentæ central, or basal. Large, tropical, land plants.
MUSACEÆ. p. 329.
Aquatic plants. Flowers usually imperfect. Ovary one to six
celled. Placentæ parietal. Style very short. Stigmas three or
six. Stamens in one or several whorls.
HYDROCHARIDACEÆ. p. 321.

119.—Anther one celled. Filament petal-like. Rhizomes usually
abounding in starch, and not aromatic. MARANTACEÆ. p. 328.
Anther two celled. Filament thread-like. Rhizomes or tubers
aromatic or resinous. ZINGIBERACEÆ. p. 328.

120.—Stamens six. **121**
Stamens three.

Anthers opening outwards. IRIDACEÆ. p. 332.
Anthers opening inwards. Seeds very minute. BURMANNIACEÆ.

121.—Fruit one celled. TACCACEÆ.
Fruit three celled.

Outer perianth green. Leaves stiff, channeled.

BROMELIACEÆ. p. 329.

Outer perianth colored.

Flowers in racemes or corymbs, borne on the stem. Stigma undivided, or nearly so. Leaves alternate, *equitant*, or folded together lengthwise. Fibrous rooted. HÆMODORACEÆ.

Leaves mostly radical, not equitant. Flowers borne on a scape, at first inclosed in a spatha-like bract, solitary or in umbels.

Leaves dry, grass-like. Seeds with a beak on the side. Embryo as long as the albumen. Roots tuberous or fibrous.

HYPOXIDACEÆ. p. 329.

Leaves succulent. Perianth imbricate in the bud.

AMARYLLIDACEÆ. p. 329.

GLUMACEOUS ENDOGENS.

122.—Stamen one. Anther one celled. Perianth consisting of a single glume, or sometimes of two nearly opposite. Leaves radical, thread-like. Australian herbs. CENTROLEPIDACEÆ.

Stamens commonly more than one. Anthers two celled.

Inner perianth petal-like. Flowers perfect, regular. Ovary formed of three carpels. Ovules numerous. XYRIDACEÆ. p. 351.

Inner perianth glumaceous.

Styles two or three to each flower. Ovary single, with a single ovule. **123**

One style to each carpel. Carpels mostly two or three, rarely single. Ovary with one ovule in each cell. Stamens two or three. Anthers mostly one celled. Flowers mostly imperfect.

RESTIACEÆ.

Style single. Ovary consisting of three carpels. Stigmas three. Stamens six or three. Anthers two celled. Flowers mostly perfect. Seeds three or more. JUNCACEÆ. p. 349.

123.—Petiole forming a split sheath which incloses a part of the stem upward from a node. Leaves somewhat widened at their junction with the sheath. Styles two, or rarely three. Embryo of the seed at the side of the albumen, near the base. Anthers versatile. Stem usually hollow, with swelled nodes. GRAMINEÆ. p. 353.

Sheath not split. Leaves not enlarged at their junction with the sheath. Embryo of the seed very small, included in the albumen at base. Filaments attached to the bases of the anthers. Stem solid, often two edged or triangular, with few nodes, and these seldom enlarged. CYPERACEÆ. p. 352.

RHIZANTHS.

124.—Ovule single, or ovules few. Ovary with one, two, or three cells. Styles one or two. BALANOPHORACEÆ. Ovules numerous.

Anthers opening by slits. Ovary one celled. CYTINACEÆ.
Anthers opening by pores. RAFFLESIACEÆ.

CRYPTOGAMIA.

The following words are much used in speaking of cryptogamous plants : —

Antheridium,— organ corresponding to the anther of a flowering plant.

Antherozoids,— cells emitted by the antheridium, which are furnished with vibratile cilia, whereby they move themselves in water, these minute cells being designed for an office similar to that of pollen.

Archegonium,— organ corresponding to the ovary of a flowering plant.

Carpospores,— spores formed after the general fertilization of the fruit-bearing organs.

Oöspores,— spores formed by direct fertilization of ovule-like cells.

Frond,—an organ which combines stem and leaf; or stem, leaf, and fruit, as in ferns.

Hypnospores,—spores that germinate after a long period of rest or sleep.

Macrospores,—spores that produce a prothallus having organs corresponding to ovaries.

Microspores,—small spores that produce a prothallus with organs corresponding to anthers.

Prothallus,—a minute plant formed by the first growth of the proper spores, and from which the ordinary plant grows after the fertilization of the archegonium.

Sporange.—a box or case in which spores are formed and come to maturity.

Spores,—proper, are commonly separate, single, minute cells which germinate to a prothallus,—though in some cases they have two or more compartments

Swarmspores,—ciliated, self-moving bodies which escape in swarms from a mother cell. Sometimes they vegetate to form new plants. Sometimes they are pollen-like in their office; and sometimes they unite two and two to form true spores or sporanges.

Thallus,—a leaf-like expansion of cellular tissue without a definite, symmetrical form.

Teleutospores,—the final spores produced by a plant which has passed through a regular cycle of generations,—the spores that are to recommence the cycle.

Zoöspores,—spores or spore-like cells capable of moving themselves by means of vibratile cilia.

Zygospores,—spores formed by the union of cells of different plants or of two zoöspores.

125.—Having roots, or hair-like organs corresponding to roots,—and a stem or a rhizome bearing leaves or branches. With rare exceptions green, and living on inorganic matter. *Acrogens.* **126**

Having no true roots, no central axis of growth, and no plain distinction of stem and leaves. Sometimes existing through life as a single cell. Commonly consisting of variously-shaped expansions of mere cellular tissue, with no fibro-vascular bundles,—the *thallus.* Simple or branched. In some cases bright green. More commonly red, purple, brown, dull green, or colorless.

Plant parasitic or growing out of ready-formed organic matter. Destitute of green chlorophyll or any other generally diffused coloring matter. Organs of fructification, or their supports, more conspicuous than the organs of vegetation, which usually consist of a *mycelium*, or loose mass of interlacing, thread-like cells. Growing on or in other plants, or in damp places, and commonly of rapid growth. Some of the minute forms are amphibious and constitute ferments. *Fungi*. **182**

Plants not parasitic, and not growing in the dark, but containing chlorophyll or green matter, so that they live on aerial food.

Thallus made up of large green or otherwise colored cells and long, thread-like, colorless cells. Growing in damp or dry places, and of slow growth. *Lichenes*. **177**

Cells of the thallus not of unlike character, or else growing in water.

Growing mostly on the earth, on tree trunks, or rocks. If aquatic, floating on the water and not-immersed. *Hepaticæ*. **148**

Growing mostly immersed in salt or fresh water. In some cases microscopic and consisting of few cells. *Algæ*. **154**

126.—Not entirely made up of cells, but having some fibro-vascular tissue. **127**

Plants consisting of a few cells, which are sometimes very long, arranged like a central stem with whorled branches. Growing immersed in fresh water, and attached to the bottom by long cells, which perform the office of roots. Spores produced in small, axillary, twisted, nut-like cases. The cells are transparent, and with the microscope colorless globules may be seen in constant circulation from end to end of each cell. These plants are usually included among the algæ. CHARACEÆ.

Consisting of cellular tissue only. Spore cases consisting of capsules furnished outside, with a calyptra. *Muscineæ*.

Calyptra separating from the capsule at its apex, and, at maturity, remaining attached to the base of the capsule or of its stalk. Capsule not furnished with a distinct lid. Spores in the capsule usually mixed with hygroscopic, spiral, elastic threads called *elaters*. Rhizoids or root hairs one celled. **148**

Calyptra separating from the capsule at its base so as to remain loosely attached like a hood or extinguisher. Capsule commonly, though not in all the sub-orders, covered by a plain lid at top. Spores never mixed with elaters. Rhizoids many celled.

Musci. **131**

(These are the true or Leafy Mosses. The so-called Scale Mosses are Hepaticæ; the Rock Mosses are Lichens; the Club Mosses belong to the Lycopodiaceæ; and the "Southern Moss" is a flowering plant, one of the Bromeliaceæ.)

127.—Stem hollow, striated, jointed, with a toothed sheath at each joint. Leafless, but usually having leaf-like branches springing in whorls from the joints. Spore cases attached to the inner side of stalked scales, which are arranged in cone-like heads at the summit of the fertile stem. * EQUISETACEÆ. p. 359.

Stem not hollow nor jointed.

Spore cases on the back or the margins of the leaf, or frond; sometimes occupying a whole frond, and sometimes a special portion of a frond or of its subdivisions, and then having a flat, leaf-like arrangement. *Filices.* **129**

Spore cases distinct from the leaves, and not having a flat, leaf-like arrangement.

Spore cases large, and situated at the bases of the leaves or leaf stalks, but not attached directly to them. Leaves distinct or none. **128**

Spore cases very small, and attached to the leaves at their axils, or to bracts which are arranged in spikes at the end of the stem.

Plants growing in the water, with no stem. Spore cases inserted in the hollowed out, enlarged bases of the long, rigid, sharp-pointed leaves. ISOETACEÆ.

Plants not growing in the water.

Having spores of one kind only. † LYCOPODIACEÆ.

Having spores of two kinds, macrospores and microspores.

SELAGINELLACEÆ.

---

* Among the fossil cryptogams the order Calamarieæ is now extinct. The plants of this order were allied to the Equisetaceæ, but they had whorls of leaves in place of the sheaths of the Equisetums.

† Of cryptogamic orders occurring in former geological epochs but now extinct; the Sigillarieæ and Lepidodendreæ were closely allied to the Lycopodiaceæ.

128.—Plants floating on the water. Leaves opposite, or else alternate and imbricate, without breathing pores. SALVINIACEÆ.
Growing in wet places, with creeping, thread-like stems. Petioles long. Leaves neither imbricated nor opposite, either having no blade, or consisting each of four heart-shaped leaflets, furnished with breathing pores. MARSILIACEÆ.

FILICES OR FERNS.

In studying the ferns the spore cases must be examined with the microscope to determine the presence and extent or the absence of a jointed, elastic ring.
Indusium,— the membranous awning over a cluster of spore cases.
Sori,— patches of fruit, forming dots or lines on the under surface or margin of the leaves.

129.—Spore cases furnished with a nearly or quite complete elastic ring. Frond circinate or unrolling endwise from the base. **130**
Spore cases with a very short ring or none.

Fronds straight, never circinate. Clusters of spore cases forming simple or pinnate spikes distinct from the leafy part of the frond which is lower down on the stem. Spore cases opening down the side. OPHIOGLOSSACEÆ. p. 563.
Fronds unrolling endwise from the bud.

Natives of tropical regions. Sori on the back of the green frond. Spore cases commonly joining together, opening by a pore, or a slit down the side. MARATTIACEÆ.
Natives of temperate regions. Sporanges often occupying entire fronds or pinnæ, so that the fertile fronds or pinnæ are unlike the green leaves. Spore cases opening across the apex.
OSMUNDACEÆ. p. 362.

130.— Ring covering the spore case like a lid or cap. Spore case opening down the side. SCHIZÆACEÆ. p. 362.
Ring vertical, extending from the base quite around the case which, when ripe, opens in a plane parallel with the base. Sori on the back or margin of the frond. POLYPODIACEÆ. p. 360.
'Ring transverse or parallel with the base of the spore case, or slightly oblique.

Ring short, extending but partly around the case, or entirely lacking. Spore case two valved, opening across the apex. Growing in temperate regions. OSMUNDACEÆ. p. 362.
Ring broad, complete. Growing in warm regions. Small and delicate.

Indusium none. Sori on the back of the leaf within the margin.
GLEICHENIACEÆ.
Indusium two valved, or urceolate. Sori at the end of a leaf vein.
* HYMENOPHYLLACEÆ. p. 362.

MUSCI. TRUE OR LEAFY MOSSES.

Apophysis,— an enlargement of the foot stalk which bears the capsule.

Calyptra, — the loose hood or covering of the capsule. It is called *conical* when it sits directly on the capsule, and is not split on one side, nor enlarged at base. It is called *hood-form* when it is split at base on one side, and rests on the apex of the capsule in a slanting position. It is called *mitre-form* when it is enlarged or inflated at base. It is said to be of the *extinguisher-form* when the base extends below the base of the capsule.

Columella,— a central stem or column within the capsule, around which the spores are formed.

Operculum,— the lid, cap, or cover of the capsule. This forms a part of the capsule itself, and must be distinguished from the calyptra, which is a loose covering but not a proper cover.

Peristome,— a fringe or membrane that is seen around the mouth of the capsule after removing the operculum.

131.— Capsule opening by a distinct lid, called the *operculum.*
Stegocarps. **133**
Capsule entire, without a distinct lid. *Cleistocarps.*

Capsule opening by longitudinal slits, the sides forming four or six valves, which remain connected at top and bottom. Leaves brown or nearly black, fleshy. ANDRŒACEÆ.

---

* Among the fossil ferns, Schimper makes out the following orders, all of which are now extinct:—
Sphænopterideæ, Neuropterideæ, Pecopterideæ, Tæniopterideæ, and Dictyopterideæ.

Capsule opening by irregular transverse rents.

Capsule globular, not pointed, sessile at the end of the stem. Spores large, few, many sided, smooth. ARCHIDIACEÆ.
Capsule pointed.

Capsule oval or oblong, with a long beak. New plants formed by offshoots springing from near the ends of the stems. 132
Capsule nearly globular. Plants very small, unbranched, annual.

Calyptra, conical, bell-form, perfect. Leaves of loose texture throughout. PHYSCOMITRIOIDEÆ.
Calyptra hood-form. Leaves of closer texture, and greener at the apex than at the base. POTTIOIDEÆ.

132.— Calyptra large, beaked, extending below the capsule, ragged at base. Alpine or Arctic·plants, throwing out rootlets all along the stem, forming branches and branchlets, and growing in dense broad tufts. Leaves wide, concave, ribbed. VOITIACEÆ.
Calyptra small, short. Leaves mostly narrow and composed of small, green cells in the narrower part, and large, colorless cells in the wider part. BRUCHIACEÆ.

133.— Pedicels and capsules growing from the sides of the stem.
Pleurocarps. 144
Pedicels and capsules growing from the end of the stem or from the ends of branches. *Acrocarps.*

Cellular tissue in the stem composed of layers of three different kinds. Capsule nearly globular. Lid hemispherical, not pointed, very small. Calyptra small and disappearing early. Spores four sided. Leaves very pale green, or colorless, not ribbed. Plants growing in wet places. SPHAGNACEÆ.
Cellular tissue nearly homogeneous. Capsules mostly elongated, having a pointed lid. Mouth of the capsule commonly fringed. Spores spherical.

Leaves with wings and a conspicuous vertical plate on the back, two ranked. 143
Leaves without a wide vertical plate on the back.

Leaves two ranked or three ranked, becoming larger and hairy towards the apex of the stem. Mouth of the capsule provided with a fringe, or peristome, visible with the microscope, when the lid is removed. Peristome composed of sixteen two-cleft, linear, or thread-like teeth. POTTIACEÆ.
Leaves two ranked, sickle-form, vertical, slightly clasping at base, very smooth and shiny. Capsule lid obtuse. Tropical plants.
DREPANOPHYLLACEÆ.
Leaves turning in every direction, or else turning the same way and sickle-form.

Peristome formed of a single membrane. WEISSIACEÆ.
Peristome formed of a single fringe. **136**
Peristome formed of a double fringe. **141**
Peristome none.

Calyptra hood-form, or else mitre-form and four sided. **134**
Calyptra either mitre-form, but not four sided, or else extinguisher-form.
Calyptra large, partly or wholly covering the capsule. Leaves of firm texture. Stem throwing out rootlets all along. GRIMMIACEÆ.
Calyptra minute, covering only the lid of the capsule. Whole plant of loose texture, rarely branching, rooting only at base.
SCHISTOSTEGACEÆ.

134.— Leaves ovate lanceolate, sometimes ragged near the apex. Cellular tissue soft. Capsule globular or pear-form. **135**
Leaves lance linear, usually channeled. Cellular tissue firm.

Capsule striated or channeled. Leaves hairy. GRIMMIACEÆ.
Capsule not striated. Leaves not hairy. WEISSIACEÆ.

135.— Leaves not ribbed, edges ragged. GRIMMIACEÆ.
Leaves ribbed.

Capsule with a distinct neck below. Calyptra four sided.
PHYSCOMITRIACEÆ.

Capsule with no distinct neck. Calyptra hood-form. POTTIACEÆ.

136.— Calyptra conical or hood-form. **137**
Calyptra mitre-form or extinguisher-form.

Peristome consisting of four three-sided teeth. Calyptra striated.
TETRAPHIDACEÆ.
Peristome consisting of sixteen teeth, which are commonly two or three cleft. Calyptra smooth, or but slightly hairy, or with an uneven surface at the apex. GRIMMIACEÆ.
Peristome consisting of thirty-two or sixty-four teeth.

Capsule resting on an enlargement of the pedicel, or apophysis, very much larger and of looser texture than the capsule itself. Calyptra very small. Leaves of loose texture. SPLACHNACEÆ.
Apophysis smaller than the capsule, and of firm texture. Calyptra furnished with long hairs, which often extend below the base of the capsule. Leaves remarkably firm in texture.
POLYTRICHACEÆ.

137.— Capsule resting on a conspicuous apophysis, larger than itself, and usually of a different color. SPLACHNACEÆ.
Apophysis none.

Plants floating on the water. Leaves ribbed. Having, at the base of the peristome, a membrane which is perforated with large openings. GRIMMIACEÆ.
Plants growing on land. Membrane at the base of the peristome, teeth none, or without openings.

Capsule oval or cylindrical. **138**
Capsule nearly globular.

Teeth of the peristome perforated with many holes. GRIMMIACEÆ.
Teeth entire or sometimes coarsely notched, but not perforated.

Capsule having a wide mouth, twisted. SELIGERIACEÆ.
Capsule globular, with a 'narrow mouth. BRYACEÆ.

138.— Teeth of the peristome sixteen, two cleft, with the two limbs usually unequal, without a common membrane at base. **139**
Teeth thirty-two or sixty-four. **140**
Teeth irregular, cut short, or ragged.

Leaves consisting of large, loose cells, not hairy.   Peristome teeth
irregular, lanceolate, two cleft, often cut short.   POTTIACEÆ.
Leaves consisting of small, dense cells.

Peristome teeth perforated with numerous holes, or irregularly two
or three cleft.   Leaves often hairy.   GRIMMIACEÆ.
Peristome teeth ragged, often abortive.   Leaves channeled, not
hairy.   Capsule with a large mouth.   SELIGERIACEÆ.
Peristome teeth irregular.   Leaves lance linear.   Capsule with a
mouth of moderate size.   WEISSIACEÆ.

139.—Leaves blue-green when moist, white when dry, very spongy.
LEUCOBRYACEÆ.
Leaves grass-green or yellow.

Teeth of the peristome undivided at the apex, but with a wide crack
or cleft in the middle of the base and extending half way up.
Calyptra open on one side almost the whole length, somewhat
twisted.   Leaves without ribs. ˙Plants small and almost stem-
less.   DISCELIACEÆ.
Teeth of the peristome purple.

Teeth long, divided to the base into two round, thread-like limbs,
remotely jointed.   CERATODONTACEÆ.
Teeth closely jointed, usually divided half way down—but sometimes to the base—into two unequal limbs, the narrower limbs
of adjacent teeth standing next to each other.   Leaves ribbed.
WEISSIACEÆ.

140.—Teeth thirty-two, thread-like, arranged in pairs.   GRIMMIACEÆ.
Teeth thirty-two or sixty-four, lanceolate, joined to the columella.
Calyptra with a roughened surface, or slightly feathery at the
apex.   POLYTRICHACEÆ.

141.—Inner peristome formed of a keeled membrane, with or without
fine hairs.   **142**
Teeth of peristome without a common membrane at base,   Inner
peristome composed of fine hairs.

Calyptra mitre-form or hood-form. GRIMMIACEÆ.

Calyptra four sided when young, enveloping a globular or pear-shaped capsule with a thick neck. PHYSCOMITRIACEÆ.

142.—Calyptra extinguisher-form, longer than the capsule.

GRIMMIACEÆ.

Calyptra hood-form or conical.

Plants without stems, or with stems hardly apparent. Leaves rare or none. Capsule large, oblique, swelled out in the middle, sessile, or with a short, thick pedicel. BUXBAUMIACEÆ.

Stem distinct, leafy. Capsule on a long pedicel. BRYACEÆ.

143.—Leaves folded together from the base, the wings being prolonged into a single vertical plate with a rib extending to the apex of this plate, or beyond. Peristome single, with sixteen unequally two-cleft teeth. FISSIDENTACEÆ.

Vertical plate formed by an extension of the rib, which is narrow at first but wider towards the apex. Wings partly folded.

EUSTICHIACEÆ.

PLEUROCARPS.

144.—Peristome double. **145**

Peristome consisting of a single fringe.

Leaves narrow, ragged, made up of large cells. Teeth of the peristome large, in pairs when young. Calyptra smooth.

FABRONIACEÆ.

Leaves wide, entire, with narrow cells.

Calyptra hood-form. NECKERACEÆ.

Calyptra mitre-form. Leaves two ranked, clasping at base. Tropical plants, but growing at high elevations.

PHYLLOGONIACEÆ.

145.—Floating in water. Branches thread-like and naked. Leaves having a keel. Capsule nearly sessile. Outer peristome of sixteen very long teeth, which are red and point inwards. Inner peristome of sixteen interlacing threads. FONTINALACE.E.
Growing on land. Leaves more or less concave, with no keel. Capsule having a pedicel. Peristome toothed, with or without threads.

Leaves arranged in two opposite ranks, with a third, stipule-like, intermediate series on the lower side of the stem. Plants growing in equatorial or southern latitudes. HYPOPTERYGIACE.E.
Leaves arranged spirally, or in many ranks.

Capsule curved. Pedicel long and bent. **146**
Capsule straight. Pedicel of moderate length, fleshy.

Stem fleshy. Calyptra conical and mitre-form, lobed at base. Pedicel thick. Capsule black. HOOKERIACE.E.
Stem slender. Calyptra hood-like. Pedicel slender. Capsule brown. NECKERACE.E.

146.—Calyptra mitre-form. Pedicel very short. NECKERACE.E.
Calyptra hood-form.

Inner peristome composed of divisions with keels, and having a common membrane at base. **147**
Inner peristome made up of threads, without a common membrane at base.

Outer teeth of the peristome lanceolate; inner hairs remote, shorter than the teeth. Cellular tissue loose. FABRONIACE.E.
Outer teeth of the peristome awl-shaped, having keels; inner hairs of about the same length as the teeth. Cellular tissue firm.
NECKERACE.E.

147.—Leaves with a roughened surface, deep green or black.
LESKEACE.E.
Leaves not roughened, green or yellow, often silky. HYPNACE.E.

HEPATICÆ.

Amphigastria,—stipule-like bracts or scales.
Elaters,—elastic spiral threads formed in the capsules among the spores.
Perianth,—a calyx-like involucrum surrounding the archegonium or fruit-producing organ or capsule.

148.—Capsules sessile, or clustered on a common pedicel. Frond without distinction of stem and leaves. **152**
Capsules solitary, on pedicels, bursting lengthwise by four valves which do not remain coherent at top.

Without distinction of stem and leaves, but consisting of a lobed or many-cleft frond. **151**
With stem and leaves distinct.

Leaves with their upper edges overlapping the lower edges of those next above. **150**
Leaves with their lower edges overlapping the upper edges of those next below.

Leaves more or less deeply lobed, often with deep incisions and narrow lobes. **149**
Leaves not lobed, but entire or simply toothed.

With no stipule-like bracts, or amphigastria, among the leaves, or having them only on the branches. Perianth free, herbaceous.
JUNGERMANNIACEÆ.
Amphigastria none, or triangular. Leaves rounded. Perianth joined at base to the involucrum. GYMNOMITRIACEÆ.
Having amphigastria on both stem and branches.

Perianth none. Fruit usually contained in sacs. JUNGERMANNIACEÆ.
Perianth free, herbaceous, or membranous. GEOCALYCACEÆ.

149.—Perianth free, herbaceous. JUNGERMANNIACEÆ.
Perianth none, or joined at base with the involucrum.
GYMNOMITRIACEÆ.

150.—Leaves entire or slightly toothed. Amphigastria forked.
CALYPOGEIACEÆ.

Leaves and amphigastria two lobed, or with hairy teeth.

Leaves with hairy teeth, or deeply two or three cleft, the divisions having long, acute points. PTILIDIACEÆ.
Leaves with two unequal lobes.

Perianth flattened, two lipped. Elaters with two spiral fibres.
MODOTIIECACEÆ.
Perianth regular or nearly so, round or angular divisions sharp pointed. Elaters with a single spiral thread. FRULLANIACEÆ.

151.—Divisions of the frond lobes resembling leaves. Perianth partly joined to the many-leaved involucrum. FOSSOMBRONIACEÆ.
Divisions of the frond not leaf-like.

Lobes of the frond without ribs. Fruit inserted in the lower face of the frond. ANEURACEÆ.
Lobes of the frond with plain ribs.

Fruit inserted on the lower face and on the ribs of the frond. Frond with elongated, forked divisions. Perianth none.
METZGERIACEÆ.
Fruit inserted on the upper face of the frond.

Perianth tubular. BLYTTIACEÆ.
Perianth none. PELLIACEÆ.

152.—Capsules arranged on a common, columnar receptacle,— two valved. Perianth none. Spores with imperfect elaters or none.
ANTIOCERACEÆ.
Capsules arranged under an enlarged, hat-like, common receptacle, which is supported by a peduncle. Capsule opening by a single slit. Spores with elaters. **153**
Capsules sessile or immersed in the frond. Spores without elaters.

Frond with a membranous wing and spirally undulated,— ribbed.
DURIŒACEÆ.
Frond more or less flattened, without a membranous wing.

Fruit sessile on the upper face of the frond. CORSINIACEÆ.
Fruit immersed in the frond. RICCIACEÆ.

153.—Involucra more or less joined together and to the receptacle, each usually containing several capsules. MARCHANTIACEÆ.

Involucra cohering at base only, having each but one capsule. Frond channeled. LUNULARIACEÆ.

Involucra at the extremities of the receptacle lobes, two valved. Capsule nearly sessile, solitary. TARGIONIACEÆ.

ALGÆ OR CELLULAR WATER PLANTS.

Brood spores,— vegetating or fertilizing cells produced within a mother cell, but not furnished with cilia, and hence not self moving.

Carpospores,— spores formed indirectly, after the fertilization of the fruit-bearing organ.

Cystocarp,— pouch or cavity containing fruit.

Cytioderm,— skeleton of mineral matter giving rigidity to the outer part of diatoms.

Cytioplasm,— colored matter contained within the cells of diatoms, and active in producing growth.

Fruit spores,— spores which result from the fertilization of the fruit-producing organ.

Oöspores,— spores produced by the direct transformation of the contents of the fruit-bearing organ, after fertilization.

Phycochrome,— matter performing the office of chlorophyll, but of a yellowish-brown or blue-green color.

Placenta,— part to which spore-bearing threads are attached.

Swarmspores,— vegetating, conjugating, or fertilizing cells which are provided with cilia, and hence are capable of moving themselves about, after escaping from the mother cell in which they are formed.

Tetraspores,— spores produced by simple division, without the fertilizing influence of other cells.

Zygospores,— spores resulting from the conjugation of two cells of apparently the same kind.

154.—Not known to increase by sexual union, but multiplying by cell division, or by swarm cells, or brood cells, or special, single, spore-like cells. Mostly microscopic plants, consisting of single cells, or of cells united end to end, so as to form simple or branch-

68

ing threads. The single-celled plants often associated in thread-
like or variously-shaped clusters, and contained in a common
mass of mucus. Chiefly fresh-water plants. **172**
Multiplying by means of zygospores, which result from the conjugation of cells of apparently the same kind. Mostly minute plants
of a bright green or greenish-brown color. **169**
Multiplying by the union of cells of different kinds to form oöspores
or carpospores.

Fruit-producing body consisting of a simple cell whose contents,
after fertilization, are directly transformed into a single spore or
many spores. Plants commonly green or olive colored. **163**
Cystocarps, after fertilization, producing in the first place special
cells, in some of which the spores are subsequently formed.
Most of the species also produce tetraspores for vegetative multiplication. Plants red or purple, rarely green or yellowish, mostly
growing in deep-sea water or below the line of low tide.

FLORIDEÆ or RHODOPHYCEÆ.

Fruit spores not in groups, but scattered singly over the frond.
Tetraspores in fours. Some growing in fresh water, some marine.
Small, thread-like or leaf-like. PORPHYRACEÆ.
Fruit spores in groups or in cystocarps.

Frond continuous, or not jointed throughout. **156**
Frond jointed.

Frond becoming hard and coral-like, with an incrustation of carbonate of lime, fruit bodies immersed in the frond, or wart-like, or
external. Frond thread-like, round, or somewhat flattened. Fruit
spores produced in fours. CORALLINACEÆ.
Frond cartilaginous, or not gelatinous. **155**
Frond consisting partly of gelatinous matter.

Frond consisting of a single stem-like row of cells with or without a
cortical layer of longitudinal threads, with dense whorls of bead-
like branches. Very slimy. Mostly growing in fresh water.
BATRACHOSPERMEÆ.

Frond having a cellular axis with surrounding bundles of longitudinal threads lying somewhat loose in the enveloping mass of mucus. Fruit spores arising from bead-like, jointed threads, at length few, rounded, angular, somewhat large, grouped in no special order. Tetraspores formed in the ends of the horizontal threads, oblong, divided by zones. Growing only in salt water. Species few. DUDRESNAYACEÆ.

155. Fruit spores formed in one or many mother cells. Fruit naked, attached sidewise to the branches. Spores numerous, rounded, angular, grouped in no special order. Frond consisting of a single row of axial cells which are sometimes covered by a layer of cortical cells. Tetraspores variously divided. CERAMIACEÆ.

Fruit spores formed in the terminal joints of articulate threads which are sometimes scattered singly, but are most commonly in groups, pear-form. Frond naked, or having a cortical layer of cells, sometimes partially continuous, but then traversed by a jointed axis. Tetraspores divided triangularly. WRANGELIACEÆ.

Spore groups contained in an external pericarp.

Spore-bearing threads free, or radiating from a basal placenta. Pericarp with an open mouth. Frond sometimes naked, sometimes with a cortical layer of cells. Tetraspores evolved among the cells around the axis of the frond, triangularly divided, definitely arranged. A very large order. RHODOMELACEÆ.

Spore-bearing threads arranged around a central placentar column, radiating upwards, mixed with sterile threads. Tetraspores on the branchlets, near the surface, divided triangularly

SPYRIDIACEÆ.

156.— Fresh water plants. Growing in fast-running water. Frond thread-like, tubular, olive colored. LEMANEACEÆ.
Growing in salt water.

Fruit body consisting of tufted spore threads. Spores formed in each cell of the spore thread, or only in the upper cells, or in the terminal cells. **157**

Fruit body consisting of one or many mother cells, in which the spores are formed by division, and remain surrounded by the transparent membrane or wall of the mother cell, in no special order.

Fruit body simple, immersed in the frond. Spores numerous.
Frond consisting of long threads, covered without by a continuous stratum of vertical threads or cells. CRYPTONEMIACEÆ.
Fruit body evidently compound, or consisting of many smaller
groups arranged in no special order. Smaller groups consisting
of few spores.

Fruit body either immersed in the frond, or contained in an external pericarp. Tetraspores in sori within the frond, or in external warts made up of threads. Divided crosswise or in zones.
Frond fleshy or gelatinous, made up of thread-like or shorter,
rounded cells. GIGARTINEÆ.
Fruit body in a pod-like receptacle. Spores numerous, rather
large. Tetraspores produced in similar receptacles, pear-form,
divided in zones. Frond fleshy, made up of interior long threads,
and exterior shorter cells. Containing only one genus.
FURCELLARIEÆ.

157.— Fruit body immersed in the frond. **160**
Fruit body contained in an external pericarp. **158**
Fruit body naked, lying in spongy or wart-like excrescences of the
frond.

Fruit bodies numerous in each excrescence of the frond, surrounded
by loose threads. Spore threads obconical, very short, radiating
from a cellular, central placenta, having single spores in the terminal joints. Frond with a rounded stem made up of long
threads, surrounded by larger, short, granular cells, and having
an external covering of small cells united into vertical, bead-like
threads. SPONGIOCARPEÆ.
Fruit bodies consisting of transformed, bead-like threads of the
warts. Spores produced in each joint of the threads, rounded or
discoid, one in each joint. Tetraspores produced by the transformation of joints of similar threads, divided crosswise or in
zones. Frond globose, or expanded horizontally. Cells held
together by mucus more or less consistent. SQUAMARIACEÆ.

158.—Placenta parietal, that is the spore-bearing threads arising
from the walls of the pericarp, and converging towards the centre. Spore-bearing threads jointed, branching, bearing simple

spores in the terminal cells of the branches. Tetraspores divided in zones. Frond tubular or very compact, made up of elongated threads or shorter cells. Pericarp at length having an open mouth. CHÆTANGIACEÆ.

Placenta basal. **159**

Placenta central.

Spore threads radiating from a central cell. Tetraspores appearing in somewhat definite patches or sori, rather large, rounded, divided triangularly. Frond flat, leaf-like, sometimes ribbed and veined, cells of the veins elongated. DELESSERIACEÆ.

Placenta in the middle of the fruit, and apparently central. Fruit spores obconical, held together by a web of very fine threads. Tetraspores immersed in the frond, divided triangularly. Frond tubular in the upper part and subdivided by cellular diaphragms. Tube surrounded by a cellular, cortical stratum.

LOMENTARIACEÆ.

159.— Fruit within the pericarp surrounded by a net work of very fine anastomosing threads. Spore-bearing threads jointed and anastomosing so as to form a network, producing rounded or angular spores in the joints. Tetraspores divided crosswise or triangularly. Frond tubular, divided by diaphragms, or close and cellular within, often with the outer cells joined into vertical threads.

CHAMPIEÆ.

Spore-bearing threads commonly branching, but not anastomosing to form a net work.

Pericarp not having a distinct mouth, one celled or two celled, Spore-bearing threads in very dense rows, having single spores in the terminal joints of the branches. Tetraspores produced among the cortical threads of the frond, divided crosswise. Frond made up of cells, or of fibres very densely interwoven. Pericarp half immersed. GELIDIACEÆ.

Pericarp usually having a distinct mouth.

Threads of the fruit much branched, many of them sterile. Upper branches of the fruit-bearing threads ripening earlier than the lower. Fruit more or less lobed, each lobe producing numerous spores grouped in no special order, in mucus. Tetraspores scat-

tered through the frond or produced in special branches. Frond sometimes tubular, but more commonly of a compacted mass of cells, the outermost cells sometimes joined to form vertical threads. RHODYMENIACEÆ.
Sterile threads of the fruit few.   Fruit-bearing threads simple or branched, ripening at about the same time.

Fruit-bearing threads jointed, bearing spores in many of the upper joints.   Spores single or many in each joint.   Tetraspores minute, more or less oblong, scattered through the slightly changed frond, or collected in superficial or external spots with a covering of threads.   Exterior cells of the frond rounded and sometimes joined into vertical threads.   Interior threads more longitudinal, inmost sometimes thread-like.   SPHÆROCOCCOIDEÆ.
Fruit-bearing threads producing a single spore in the larger, terminal, club-shaped cell.   Tetraspores in the cortical stratum of the frond, or in warts made up of threads.   Frond tubular or solid, made up of threads or cells not arranged in straight lines.
CHONDRIEÆ.

160.—Frond hardened by an incrustation of carbonate of lime, coral-like.   Fruit spores produced by fours.   CORALLINACEÆ.
  Frond soft, flexible, seldom incrusted.

Fruit without a proper placenta, the threads lying in bundles within the frond.   Fertile threads but little branched; when young, bead-like, with very short, rounded joints.   Spores few, large, grouped in no special order, held together by mucus.   Frond often gelatinous.   DUMONTIACEÆ.

Placenta central.   **162**
Placenta basal or from the sides of a central dissepiment, parting two adjoining fruits.   **161**
Placenta parietal.   Spore-bearing threads converging towards the centre from a net work of threads next the wall.   Spores single, borne in the terminal joints of the branches of the fertile threads.

Fruit compound, the parts being arranged in no special order. Spores pear-form.   HYPNEACEÆ.
Fruit simple.   Spores club-shaped.   CHÆTANGIACEÆ.

161.—Threads of the fruit much branched, many of them sterile. Upper branches of the fruit-bearing threads ripening earlier than the lower Fruit more or less lobed, each lobe producing numerous spores grouped in no special order in mucus. Tetraspores scattered through the frond or produced in special branches. Frond sometimes tubular, but more commonly of a compacted mass of cells, the outermost cells sometimes joined to form vertical threads. RHODYMENIACEÆ.

Fruit-bearing threads ripening at about the same time. Spores single, borne in the terminal joints of the threads.

Spore-bearing threads in very dense, numerous, short bundles. Tetraspores produced among the cortical threads of the frond, divided crosswise. Frond made up of cells or threads very densely interwoven. Fruit single or joined by twos. GELIDIACEÆ.

Spore-bearing threads long and branching, radiating in bundles from the placenta, producing single spores in the larger, terminal club-shaped cells. Tetraspores in the cortical stratum of the frond, or in warts made up mostly of threads. Frond tubular or solid, composed of threads or of cells not arranged in straight lines. CHONDRIEÆ.

162.—Fertile fruit threads bearing many spores in their upper joints. Fruit compound, the parts arranged radially around the placenta. Spores contained in mucus. Tetraspores divided by zones.
ARESCHOUGIEÆ.

Spores single, in the upper joints of the fertile spore threads.

Fruit compound, the parts radially disposed around the central placenta. Spores pear-shaped or club-form, borne singly in the terminal joints of the thread branches, not contained in mucus.
SOLIERIACEÆ.

Fruit simple, with hardly any proper placenta, but made up of a bundle of spore-bearing threads. Spores borne in the upper joints of the threads. Frond often cylindrical, gelatinous, sometimes covered with a crust of carbonate of lime.
HELMINTHOCLADEÆ.

163.—Plants mostly small, of a bright green color, but occasionally red. Usually containing starch granules, with a nucleus in each cell. Coloring matter not changed by acids or alkalies. Mostly growing in fresh water. **165**

Plants mostly large, of a dark olive green color, growing in salt water.

MELANOPHYCEÆ or FUCOIDEÆ.

Frond formed of threads which are either single or united into a bundle, or a compound body. **164**

Frond leathery or membranous, making a compact cellular substance. All growing in salt water.

Fruit contained in spherical cavities in the substance of the frond. Found in all seas. FUCACEÆ.

Fruit attached to external jointed threads, which are either free or compacted into knob-like masses. Small plants found in all seas. SPOROCHNACEÆ.

Fruit attached to the outer surface of the frond.

Sporanges arranged in patches of definite outline. More common in warm regions. DICTYOTACEÆ.

Sporanges scattered in patches of irregular outline, or else covering the whole surface. Plants commonly very large. Nearly all found in the northern hemisphere and in the colder waters. LAMINARIACEÆ.

164.— Frond cartilaginous or gelatinous, the interior portion being made up of thread-like, longitudinal cells, the exterior of short, jointed, radial rows of cells. Sporanges scattered over the whole frond, and immersed in its substance. CHORDARIACEÆ.

Frond jointed, simple or branching. Sporanges single, on the ends or sides of special shoots.

Body and branches of the frond made up of cells in a single row. Nearly all growing in salt water. Small and mostly parasitic. ECTOCARPACEÆ.

Frond made up of cells in several rows. Branches often ending in an enlarged cell which has been changed by a parasitic plant. Small plants growing in salt water. A few parasitic. SPHACELARIACEÆ.

CHLOROPHYLLOPHYCEÆ.

165.—Individual plants one or two celled. **168**
Individual plants composed of many cells.

Thallus consisting of threads jointed throughout. **166**
Thallus membranous or thread-like and not jointed throughout.

Thallus membranous, leaf-like, or formed into an irregular sac or
tube. Cells minute, in vegetation dividing into twos or fours.
ULVACEÆ.

Frond consisting of a simple or branched smooth thread, with
whorls of jointed branches. Having root-like projections at base.
Growing in salt water. DASYCLADEÆ.

Cells inflated. Frond consisting sometimes of one large, long cell,
sometimes of several cells cohering into a long thread or honey-
combed membrane. The threads sometimes cohere into a net
work or a leaf-like expansion. Growing only in salt water and
in warm climates. VALONIACEÆ.

166.— Threads not branched. Cells very long, cylindrical, all alike.
Chlorophyll arranged in ring-like bands. Spores formed in one
cell very numerous. SPHÆROPLEACEÆ.

Threads sometimes simple, sometimes branched. **167**
Threads always branched.

Growing in the air. Filled with red or orange coloring matter
equally distributed. Having a smell of violets.
CHROÖLEPIDACEÆ.

Growing in water. Green.

Branching by twos. Terminal cells appearing as long, colorless
hairs. Imbedded in mucus. CHÆTOPHORACEÆ.

Branching variously. Threads bent in the middle, pointed at the
ends. Genus *Asterothrix*.

167.— Cells all alike, not surrounded by mucus. Green matter, sometimes uniformly distributed, sometimes variously grouped. (The
exact place of this family is still uncertain.) CONFERVACEÆ.

Cells not all alike, showing a peculiar cap-like arrangement at the
ends during vegetative division. Cells short. ŒDOGONIACEÆ.

168.— Cells single but arranged in groups within a mass of mucus. Each cell furnished with two cilia projecting out through the mucous envelope. All the cilia move in unison, so that the group is in constant motion. VOLVOCINEÆ.
Cells not capable of locomotion. In time of fructification mostly two celled. Cells often variously branching. Growing in tufts. Sometimes the cells are associated into a spongy frond-like mass. Growing in fresh water. VAUCHERIACEÆ.
Growing in salt water.

Cells in multiplication remaining associated so as to appear like a branched thallus. VALONIACEÆ.
Single cells so branching as to imitate stem, root, and leaves. Growing very large, 10 to 15 centimetres long. CAULERPEÆ.
Thallus having root-like projections for adhesion to other bodies, much branched, with the fine branches so interlaced as to appear like a mass of cellular tissue. CODIEÆ.

169.— Union affected by ordinary vegetative cells, not furnished with cilia. **170**
Union effected by similar, self moving, ciliated swarm cells. Cell contents green.

Microscopic one-celled plants. Cells becoming pear-shaped and having numerous, root-like branches. Growing in mud, or in partially dried pools, and capable of assuming various appearances, according to the amount of moisture and light.
HYDROGASTREÆ.
Plants composed of many short cells united into threads which are cylindrical or bead-like. ULOTHRICACEÆ.
Cells single but commonly associated in clusters. Fresh-water plants.

Cells living singly or in clusters and surrounded by mucus.
PANDORINEÆ.
Cells associated in disc-like clusters or in hollow nets, not surrounded by mucus. HYDRODICTYEÆ.

170.— One-celled plants, though the cells often remain associated in thread-like clusters. **171**
Cells united into threads. Of a bright-green color.

Contents of both conjugating cells uniting in one of the original cells. ZYGNEMACEÆ.
Contents of the conjugating cells uniting in an intermediate cell.
MESOCARPEÆ.

171.—Of various forms, but mostly with a central contraction forming symmetrical halves. Cell contents bright green. Wholly destroyed by a strong heat or by strong mineral acids. DESMIDIACEÆ.
Cell contents yellowish or orange. Cytioderm or cell skeleton consisting mainly of silica, so that the form and the beautiful characteristic markings remain after the decay of the cell contents, or their destruction by fire or acids, but completely dissolved by caustic alkalies. Cell contents turned green by death or by acids, but not changed by alkalies. DIATOMACEÆ.
Rabenhorst's classification of the Diatomaceæ, Diatomophyceæ, or Bacillariæ is given in **174.**

172.—Microscopic, one-celled plants of a bright-green color, but occasionally red or orange. Cells often associated in clusters so as to appear like a single body. Cells or cell clusters frequently contained in mucus. Mostly growing in fresh water or on wet substances. **173**
Cells of a dull green, blue-green, or brown color, single or associated in threads or variously-shaped clusters. Some species have an unpleasant odor. *Cyanophyceæ* or *Phycochromophyceæ*.

Truly branching. SIROSIPHONIACEÆ.
Not branching, or only apparently branching; but sometimes spreading from a common sheath.

One-celled plants. Cells single, or many associated in clusters, and contained in a common covering, surrounded by mucous matter.
CHROÖCOCCACEÆ.
Many-celled plants. Cells divided in only one direction.

Vegetation limited at the ends of the threads. Threads jointed, slender at the upper extremity, often drawn out to a long thread-like point; provided at base with a persistent vegetative cell.
RIVULARIACEÆ.
Vegetation not limited at the ends.

Spuriously branched. Threads or trichomes tufted, not surrounded
by slime, but sheathed, provided with interstitial cells. Sheaths
firm, in layers. SCYTENEMEACEÆ.
No apparent branching.

Threads more or less distinctly jointed, sheathed, self moving.
OSCILLARIACEÆ.
Threads more or less bead-like, not moving, with permanent cells.
NOSTOCHACEÆ.

173.— Increasing by the formation of free cells within a mother cell.
PROTOCOCCACEÆ.
Increasing by cell division. Cells contained in mucus, either single
or in clusters. PALMELLACEÆ.

SUB-ORDERS OF DIATOMOPHYCEÆ.

174.— Provided with radiating spines. Marine or fossil. ACTINISCEÆ.
Without sharp-pointed projections,— or pointed only at the two
extremities.

Rarely wedge shaped. **175**
Wedge shaped, but not bent.

With central nodules in front view, and longitudinal lines in side
view. Growing in fresh water. GOMPHONEMEÆ.
With no central nodules in either view.

Transverse striæ interrupted by an open space. SURIRELLEÆ.
Transverse striæ either very short, or not interrupted by an open
space. Mostly marine. MERIDIACEÆ.

175.— In front view sometimes nearly rectangular or rhomboidal, sometimes equilaterally triangular. Having rounded projections at
the corners. Apparently many celled. Valves joined by a band
or girdle which is very narrow in the triangular species, but in
others is very broad and conspicuous. BIDDULPHIEÆ.
In front view round. Dotted, often radially. In side view appearing as flat or convex discs. MELOSIREÆ.
Not circular and not having rounded projections at the corners.

Without central nodules. **176**

With a central nodule or with two or more nodules away from the centre.

In side view appearing as if bent. ACHNANTHEÆ.

In side view, unsymmetrical, with one edge convex, the other nearly straight. CYMBELLEÆ.

Symmetrical in both front and side views.

In side view rectangular. In front view long, with a nodular inflation at each end as well as in the centre. TABELLARIEÆ.

In side view rounded or truncate at the ends. In front view boat shaped; with a central, longitudinal line; usually with small, not inflated, terminal nodules. Sometimes in place of the central nodule there is a transverse space free from striæ. NAVICULACEÆ.

In side view rectangular. In front view oval or elliptical. FRAGILARIEÆ.

176.— With longitudinal or marginal keels. **176, b**
Without keels.

In front view somewhat crescent shaped. Uninterruptedly ribbed, striated, or dotted. EUNOTIEÆ.

In front view oval, elliptical, or nearly circular. In side view sometimes rectangular, sometimes flexuous. With radiating, or interrupted, transverse ribs, often with the intervals dotted. SURIRELLEÆ.

In front view elliptical or oval, sometimes attenuated at the ends. Smooth, ribbed, or interruptedly striated transversely. In side view rectangular, linear. FRAGILARIEÆ.

176, b.— In front view oval or elliptical, sometimes contracted at the central part. Edge furnished with a keel. Transversely striated or ribbed. SURIRELLEÆ.

Not transversely striated. With one to three longitudinal keels or ribs. Cylindrical or spindel form. AMPHIPLEUREÆ.

Transversely ribbed. Ribs very short, or not continuous. Keel usually nearer one margin than the other. Pointed at the ends in one view and generally in both. NITZSCHIEÆ.

LICHENS.

Apothecium,— an open fruit patch.

Crustaceous,— forming a hard, dry incrustation, closely adhering to the surface on which it grows.

Gonidia,— large, green, gray-green, or yellow cells, single or in thread-like clusters contained in the net work of fine, thread-like, colorless cells.

Hypothallus,— a membrane or substratum consisting of thread-like cells, but differing in appearance from the rest of the plant.

Perithecium,— an envelope or involucrum surrounding a narrow-mouthed fruit patch.

Stroma,— a thick mass of thread-like cells.

177.— Apothecium circular, spread out flat, with a margin of the same structure as the thallus itself, that is, colored by green or grayish gonidia. *Parmeliacei.* **181**
Fruit patches usually with a border or involucrum unlike the rest of the thallus, or else without a border.

Perithecium globular, pierced by a small opening at the apex.
Verrucariei. **179**
Fruit patch in an open apothecium.

Apothecium rounded, spread out flat, plate-form.
Lecidiacei. **180**
Apothecium bowl-form or globular, mostly borne on a short stalk, its disc closely covered with naked spores. *Caliciacei.* **178**
Apothecia of various forms, but usually elongated and furrow-like. Thallus crustaceous, or like a dry incrustation. *Graphidiacei.*

Apothecia long, narrow, and furrow-like. OPEGRAPHEI.
Apothecia many, collected in a stroma, or swelled out mass.
GLYPHIDIEI.
Apothecia roundish, flat, rarely elongated. LECANACTIDEI.
Apothecia partially running together, of various forms, with no border. ARTHONIEI.

178.— Thallus erect, branching, bushy. SPHÆROPHOREI.
Thallus crust like. CALICIEI.

179.— Thallus leaf-like, or scale-like. ENDOCARPEI.
Thallus crust-like. VERRUCARIEI.

180.— Thallus crust-like, adhering to its support. LECIDIEI.
Thallus horizontal, thread-like. CÆNOGONIEI.
Thallus consisting of two parts. The horizontal part scaly or crustaceous, sometimes disappearing soon; vertical part with an apparent stem. CLADONIEI.

181.— Thallus mostly thread-like, and of nearly uniform character in all parts, erect, or at length hanging down. USNEEI.
Thallus crust-like when dry. **181, b.**
Thallus leaf-like when young, and not gelatinous.

Thallus rising obliquely, variegated underneath by veins or marks. Gonidia green or bluish. PELTIGEREI.
Thallus horizontal.

Thallus furnished with fibrils on the under surface. PARMELIEI.
Thallus joined centrally to its support by a stem-like cord.
UMBILICARIEI.
Thallus at length becoming scale-like, and in some places diminished to a crust, resting on a plain hypothallus. PANNARIEI.

181, b.— Thallus gelatinous when wet, rarely thread-like, sometimes leaf-like when young, but diminishing at length to a crust. Gonidia of a peculiar character. COLLEMEI,
Thallus not gelatinous, sometimes uniform, sometimes figured.
LECONOREI.

FUNGI.

Ascus,— a minute sac usually long and club-shaped, in which several spores are formed.
Basidia,— elongated or club-shaped cells, bearing one, two, or four spores at their points.
Conidia,— cells or spores not resulting from fertilization, often formed in bead-like strings, and thrown off one by one, capable of sprouting to form new plants.

Hymenium,— a surface of greater or less extent, and variously situated, bearing basids which produce spores.

Hypha,— a thread, usually one bearing spores.

Mycelium,— the thread-like, vegetating part of fungi, making a loose felt of very long colorless threads.

Peridium,— a wrapper or envelope protecting clustered spores.

Perithecium,— a small envelope or involucrum containing spores.

Stroma,— a close body formed by the association of threads.

Sclerotium,— a hard, stem-like body produced by the induration of united threads of mycelium.

182.— Growing in the air, or in other plants or in animals. **183** Growing in water or watery liquids.

Growing on decaying vegetables or animals. Mostly one celled, but much branched. Producing vegetating swarmspores in special cells at the ends of branches. Reproducing by means of fertilized oöspores. SAPROLEGNIEI.

Sexual reproduction unknown. One-celled ferments.

Schizomycetes.

Multiplying by cell division. Very minute, microscopic organisms, often in a state of lively motion. Producing or accompanying disease and putrefaction. BACTERIACEÆ.

Multiplying by budding, or by the formation of brood cells. Producing fermentation. SACCHAROMYCETES.

183.— Multiplying by means of vegetating, ciliated swarmspores. Probably reproducing also by means of zygospores. One-celled or two-celled plants, parasitic in algæ, or in the epidermis cells of phanerogamous plants. CHYTRIDIACEÆ.

Forming zygospores by the conjugation of special cells. Producing also vegetative spores in heads, not ciliated. Molds. MUCORINI.

Not multiplying by means of ciliated swarmspores, nor reproducing by zygospores.

Spores vegetating to minute, naked, self-moving, gelatinous bodies, which unite to a creeping, slimy mass, or *plasmodium*, that at length produces spore cases. MYXOMYCETES.

Spores vegetating to a cellular mycelium.

Spores produced in *oögonia*, or cells whose contents are fertilized, by other cells called *antheridia*. , Parasitic fungi spreading their mycelium in the intercellular spaces of living plants. Producing, besides the true spores, conidia, which are successively detached like beads from the ends of branches, and are capable of vegetating directly, but sometimes form vegetating swarmspores.

PERONOSPOREÆ.

Spores produced naked on the points of spicules of larger cells called *basids*, which are sometimes exposed, and sometimes inclosed in a head. BASIDIOMYCETES. 186

Spores produced on threads. 190

Spores produced in minute sacs or asci.

Asci naked or arranged on an exposed surface. *Discomycetes.* 185
Asci contained in a head, or within a peridium. *Pyrenomycetes.*

Fruit very small, not growing underground. 184
Fruit large, growing underground.

Fruit cork-like, or woody outside, of uniform color within. Bearing, at length, a dusty mass of spores inside. ELAPHOMYCETES.
Fruit veined or marbled inside. Interior wavy and deeply folded. TUBERACEI.

184.—Fruit head entirely closed. Growing mostly on living leaves. Producing bead-like conidia as well as true spores.

PERISPORIACEI.

Fruit head with a mouth-like perforation. SPHÆRIACEI.
Fruit head opening by an irregular rupture at the top. Asci borne on branched threads which stretch across the cavity of the head in every direction. Growing mostly on animal substances.

ONYGENEI.

185.— Asci produced singly on branches of the mycelium, or without a mycelium. Mostly minute fungi, parasitic on living plants.

GYMNOASCI.

Plant made up of black, jointed threads, bearing here and there sporanges or asci. ANTENNARIEI.
Asci in groups.

Fruit bodies mostly flat, hemispherical, ellipsoidal, or long, at length bursting irregularly, and showing groups of asci only at the base. Horn-like, leathery, or membranous, mostly black and sessile, very minute. PHACIDIACEI.

Fruit bodies more or less bowl-shaped, covered with groups of asci, on the flat or concave sides. PEZIZACEÆ.

Fruit body in the form of a cap, or club-shaped, fleshy, on a stalk, coated with asci groups on the outer surface of the cap or club, not minute. ELVELLACEI.

186.—With no special hymenium or basid-bearing surface, the spores being formed on single basids, growing directly from the myce-lium. **189**

Hymenium not inclosed. **187**

Hymenium inclosed in a head. *Gasteromycetes.*

Fruit body globular, circular, or beaker-form, often opening in a star-form, or like a lid. Inner mass transforming itself into small, lens-form or globular bodies, so as to appear like eggs in a nest. NIDULARIACEI.

Inner mass of the fruit body consisting of many small, irregular cavities, on the sides of which the basids are formed.

Growing underground, the fruit body remaining closed till it decays.
HYMENOGASTREI or HYPOGÆI.

Growing above ground.

Fruit body with two coats and many cavities, the inner part being borne aloft by the rapid growth of the stalk till it appears like a cap on which the hymenium is exposed. Of unpleasant smell, and very soon decaying to a liquid mass. PHALLOIDEI.

Interior of fruit body arranged in irregular folds. At length drying to dusty spores. Having a *volva*, or envelope. Growing in warm regions. PODAXINEI.

Interior of heads cellular, drying up to a mass of threads and dusty spores. (Puff balls.) LYCOPERDACEI or TRICHOGASTRES.

187.—Hymenium on the under side. **188**

Hymenium on the upper or outer side.

Substance of fruit body gelatinous, and lobed, convoluted, or disc-like. TREMELLINI.

Substance firm in texture. Club-shaped, simple or branching.
CLAVARIEI.

188.—Fruit-bearing surface even, rough, veined, or wrinkled.

AURICULARINI.

Hymenium in the form of cylindrical or irregular tubes, appearing on the under surface like dots, pores, or irregular, vertical folds.

POLYPOREI.

Hymenium in the form of spines, prickles, teeth, or warts, projecting downwards. HYDNEI.

Hymenium in vertical plates or gills, radiating outwards from a central stalk. Umbrella-like. AGARICINI.

189.—Parasitic on the leaves of Vaccinium, and forming a yellow or red rust on the upper side of the leaf, and a white coat on the under side. Genus *Exobasidium.*

Rusts growing on and in the leaves of living plants. Each species capable of appearing under different forms, and of producing reproductive bodies of different kinds. *Uredo spores* or *Summer spores* are formed singly on the branches of basids which grow directly from threads, produced by the germination of *Winter spores*, which are made up of two cells or more. *Æcidium* spores are conidia formed in cups or receptacles on the surface of green leaves, after the sprouting of the Summer spores.

UREDINEÆ.

Brands parasitic within green leaves, producing spores and sporidia.

USTILAGINEÆ.

Parasitic in insects. Producing yeast-like cells which multiply in the fluids of the host and finally form a mycelium.

ENTOMOPHTHOREÆ.

Many of the remaining orders include plants that are not independent species, but forms or stages of other orders. But for convenience of reference they are here inserted as given by Cooke.

190.—Reproductive portion of the plant predominant. Spores borne at the end of short, inconspicuous threads, which are either naked or contained in a perithecium or peridium. *Coniomycetes.* **192**

Vegetative threads conspicuous, commonly free, or loosely associated, but sometimes compacted into a common stem. *Hyphomycetes.*

Fertile threads free or loosely interlacing. **191**

Fertile threads compacted together, sometimes cellular. Stem or stroma compound.

Spores dry. ISARIACEI.
Mass of spores moist, somewhat gelatinous. STILBACEI.

191.—Fertile threads blackened. Spores mostly compound.

DEMATIEI.

Fertile threads not blackened.

Fertile threads hardly distinct from the mycelium. Spores very abundant. SEPEDONIEI.
Fertile threads distinct from the mycelium.

Spores mostly simple, naked. MUCEDINES.
Spores covered with threads which form a kind of envelope.

TRICHODERMACEI.

192.—Growing on living plants. Described in **189** and **193**
Growing on dead or dying plants.

Growing on the outside of the plant. Fructifying surface naked.

TORULACEI.

Growing under the surface or cuticle of other plants.

Perithecium more or less distinct. SPHÆRONEMEI.
Perithecium none. MELANCONIEI.

193.—Peridium cellular. ÆCIDIACEI.
Peridium none.

Spores nearly globular, simple. CÆOMACEI.
Spores mostly oblong. usually divided into two or more compartments. PUCCINIÆI.

INDEX.

The first figures after the names of the orders refer to the numbers in the table under which the names occur.

The numbers after "sp." indicate the number of species of the order as given for dicotyledonous plants, and some monocotyledonous, in De Candolle's Prodromûs; for monocotyledonous plants in Kunth's Enumeratio; for Filices in Hooker and Baker's Synopsis Filicum; for Musci in Muller's Synopsis; for fresh water Algæ in Rabenhorst's Algæ Aquæ Dulcis; for Florideæ and Fucoideæ in Agardh's Species; for Lichens in Nylander's Lichenes; for Fungi in Cooke's works and Fries' Hymenomycetes. Of course these numbers do not accurately express the whole number of species known.

91

93